Affective Geographies of Transformation, Exploration and Adventure

Combining critical reflections from scholars around the globe as well as experiential records from some of the world's most tenacious explorers, this book interrogates the concept of the 'frontier' as a realm of transformation, exploration and adventure.

We discover the affective power of social, physical, spiritual and political frontiers in shaping humanity's abilities to change and become. We collectively unpack the enduring conceptualization of the frontier as a site of nation-state identity formation, violent colonization, masculine prowess and the triumph of progress. In its place, this book charts a more complex and subtle emotional geography amidst an array of frontiers: the expanding human psyche that is induced under free-diving narcosis and tales of survival on one of the most technically difficult mountains in the world, 'The Ogre'. Chapters consider solitude in the Sahara, near-death experiences in Tibetan Buddhism, the aftermath of a volcanic eruption in Bali, the Spanish Imaginary, snatched moments of sexual curiosity, and many more.

This book will be of upmost importance to researchers working on theories of affect, the Anthropocene, frontier theory and human geography. It will be vital supplementary reading for undergraduates and postgraduates on courses such as Heritage Studies, Human and Cultural Geography, Anthropology, Tourism Studies and History.

Hayley Saul is a lecturer at Western Sydney University, Australia, and Director of the Himalayan Exploration and Archaeological Research Team (HEART) research group. Following her Ph.D. at the University of York she undertook AHRC-funded post-doctoral research with the Early Pottery in East Asia project and was a Japan Society for the Promotion of Science post-doctoral fellow on the Japanese Archaeo-Ceramic Residue Research Strategy (JARRS) project. Her current research project, the Chaturale Museum of Cuisine, with a Nepalese community on the outskirts of Kathmandu, draws together her interests in culinary heritage and Nepalese archaeology.

Emma Waterton is Associate Professor in the Geographies of Heritage at Western Sydney University, Australia. She was a Research Councils UK (RCUK) Academic Fellow at Keele University from 2006–2010 and a DECRA Fellow at WSU from 2012–2016. Her research explores the interface between heritage, identity, memory and affect in both Australian and international contexts. She has published nineteen books, including the monographs *Politics, Policy and the Discourses of Heritage in Britain* (2010), *Heritage, Communities and Archaeology* (co-authored with Laurajane Smith; 2009) and *The Semiotics of Heritage Tourism* (co-authored with Steve Watson; 2014).

Critical Studies in Heritage, Emotion and Affect

In Memory of Professor Steve Watson (1958–2016)

Series Editors: Divya Tolia-Kelly (Durham University) and Emma Waterton (Western Sydney University)

This book series, edited by Divya P. Tolia-Kelly and Emma Waterton, is dedicated to Professor Steve Watson. Steve was a pioneer in heritage studies and was inspirational in both our personal academic trajectories. We, as three editors of the series, started this journey together, but alas we lost his magnificent scholarship and valued counsel too soon.

The series brings together a variety of new approaches to heritage as a significant affective cultural experience. Collectively, the volumes in the series provide orientation and a voice for scholars who are making distinctive progress in a field that draws from a range of disciplines, including geography, history, cultural studies, archaeology, heritage studies, public history, tourism studies, sociology and anthropology – as evidenced in the disciplinary origins of contributors to current heritage debates. The series publishes a mix of speculative and research-informed monographs and edited collections that will shape the agenda for heritage research and debate. The series engages with the concept and practice of Heritage as co-constituted through emotion and affect. The series privileges the cultural politics of emotion and affect as key categories of heritage experience. These are the registers through which the authors in the series engage with theory, methods and innovations in scholarship in the sphere of heritage studies.

Heritage, Affect and Emotion
Politics, practices and infrastructures
Edited by Divya P. Tolia-Kelly, Emma Waterton and Steve Watson

Visualising Place, Memory and the Imagined
Sarah De Nardi

Affective Geographies of Transformation, Exploration and Adventure
Rethinking Frontiers
Edited by Hayley Saul and Emma Waterton

For more information about this series, please visit: https://www.routledge.com/Critical-Studies-in-Heritage-Emotion-and-Affect/book-series/CSHEA

Affective Geographies of Transformation, Exploration and Adventure

Rethinking Frontiers

Edited by
Hayley Saul and Emma Waterton

Routledge
Taylor & Francis Group
LONDON AND NEW YORK

First published 2019
by Routledge

2 Park Square, Milton Park, Abingdon, Oxfordshire OX14 4RN
52 Vanderbilt Avenue, New York, NY 10017

Routledge is an imprint of the Taylor & Francis Group, an informa business

First issued in paperback 2020

British Library Cataloguing-in-Publication Data
A catalogue record for this book is available from the British Library

Library of Congress Cataloging in Publication Data
A catalog record has been requested for this book

ISBN: 978-1-138-70112-0 (hbk)
ISBN: 978-0-367-58918-9 (pbk)

Typeset in Times New Roman
by Taylor & Francis Books

Steve Watson, 14 June 1958–22 January 2016

Steve Watson, one of the original editors of the book series in which this collection sits, died on 22 January 2016, aged 57. In addition to the energy and spirit he gave to the series, he was instrumental to the conceptualization of this volume, which he developed with Hayley over conversations, tea and biscuits on the outskirts of York in 2015. While the collection and overarching series are clear expressions of his intellect and scholarship, we hope this dedication, though brief, may give shape to his compassion, grace, collegiality and warmth. His support was constant; his good humour endless; his scholarship flawless. He opened the gates of research for so many students and colleagues, and dedicated hours of his time to discussing, debating and coaxing fresh ideas out of those around him; his generosity in doing so has fundamentally influenced the field of Heritage Studies. He was a true scholar and an even better friend, and he continues to inspire us personally and academically. We feel privileged to have had him in our lives. This volume was conceived and shaped with Steve's lead, and completed in his memory. We dedicate it to him.

Contents

Figures

Contributors

Esther Bott is a lecturer in the School of Sociology and Social Policy at the University of Nottingham, UK. Her Ph.D., which she completed in 2007, explored consumption and identity in British migration to Tenerife. Esther's current and future research interest lies in the cultural and social impacts of 'new' forms of tourism. It revolves around the global dynamics of risk implicit in high-risk adventure tourism; the negotiation of newly emergent labour markets by locals as a result of tourism patterns; the gendered impacts of tourism development in the developing world; and the discursive reproduction of 'exotic' adventure sites.

Denis Byrne is a senior research fellow at the Institute for Cultural Studies at Western Sydney University, Australia. He is an archaeologist working in the fields of heritage conservation and heritage studies, focusing on Australia and Southeast Asia. His research interests include the heritage of transnational migration and the history of archaeological heritage management in Asia. His books include *Surface Collection: Archaeological Travels in Southeast Asia* (2007) and *Counter Heritage: Critical Perspectives on Heritage Conservation in Asia* (2014).

Jenny Hall is a mountaineer and cultural geographer studying for a Ph.D. at York St John University, UK. Her research interests focus on the interdisciplinary aspects of tourism geography, extreme tourism, affect, gender, embodiment and emotion in mountaineering. She is a member of the Royal Geographical Society, a qualified mountain leader and a member of her local Mountain Rescue Team in the North of England.

Robin Hanbury-Tenison, OBE, is an explorer and President of the charity Survival International. He has been a council member – and a gold medallist – of the Royal Geographical Society; winner of the Pio Manzu Award; an international fellow of the Explorers Club; Winston Churchill Memorial Fellow; trustee of the Ecological Foundation; fellow of the Linnean Society; and was made an officer of the Order of the British Empire (OBE) in the 1981 New Year Honours. He has authored and edited over twenty books on the topics of travel and exploration.

Rosemary Kerr is a historian and heritage consultant based in Sydney, Australia. She completed her doctorate in History at the University of Sydney in 2012. Her thesis, 'On "the Road": A Cultural History', explored imaginings, experiences and representations of the Australian road since 1890. It was shortlisted for the Australian Historical Association's Serle Award in 2014. She is currently preparing a monograph based on the thesis for publication with Channel View Publications. Rosemary is a member of the Professional Historians Association (NSW) and ICOMOS.

Tim Macartney-Snape led the first Australian expedition to Mount Everest in 1984 when his team pioneered a new route on the north face and startled the mountaineering world by making the ascent without bottled oxygen. He returned to the mountain six years later, after walking there from the Bay of Bengal, whereupon he climbed alone and again without bottled oxygen. This event was captured on film in the multi-award-winning documentary Everest from Sea to Summit. This achievement marked the culmination of twelve years of successful Himalayan ascents on ten major expeditions, all attained without bottled oxygen in a daring, lightweight style, and some up previously unclimbed routes on iconic peaks such as Gasherbrum IV, Annapurna II and Ama Dablam. The state of mind that enabled Tim to succeed on those pioneering climbs led him to found Sea to Summit, an outdoor and adventure gear manufacturer that is now a global leader in that sector. He is chair of Leave No Trace Australia and a patron of the World Transformation Movement, an organization dedicated to understanding and ameliorating the human condition. He has given inspirational talks around the world and especially enjoys sharing his experiences with young people.

Felicity Picken is a lecturer in Social Sciences at Western Sydney University, Australia. Her work contributes towards the broader scholarship concerned with changing relations between humans and nature and the strange living out of the Anthropocene. Her current book project focuses on the emergence of the 'blue planet' as a significant social actor and explores how relations with oceanic environments are evolving. When on land, she explores tourism and travel as a significant shaper of twenty-first-century life and the contributions of architecture, art, heritage and urban design in legislating novel kinds of places. She can often be found applying a critique of humanist ontologies across all of these domains in the pursuit of a more symmetrical understanding of life.

Hayley Saul is a lecturer at Western Sydney University, Australia, and Director of the Himalayan Exploration and Archaeological Research Team (HEART) research group. Her research explores the heritage and archaeology of the Himalayas, and the interface of these disciplines with development agendas. HEART operates in partnership with Nepalese communities as well as a range of charity NGOs. It aims to bring positive benefits to rural Nepalese communities by operationalizing heritage and archaeology researches. Following her Ph.D. at the University of York, Hayley undertook AHRC-funded post-doctoral research with the Early Pottery in East Asia project and was a Japan Society for the Promotion of Science post-doctoral fellow on the Japanese Archaeo-Ceramic Residue Research Strategy (JARRS) project to investigate cuisine and foodways in prehistoric hunter–gatherer groups in Denmark, northern Germany and Japan throughout the Holocene. Her current research project, the Chaturale Museum of Cuisine, with a Nepalese community on the outskirts of Kathmandu, draws together Hayley's interests in culinary heritage and Nepalese archaeology.

Douglas Keith Scott, CBE, is a British mountaineer noted for the first ascent of Everest's South-West Face on 24 September 1975. He is regarded as one of the world's leading high-altitude and big-wall climbers and has made over forty major ascents in the Himalaya and Greater Ranges. He was made a CBE in 1994, and in 1999 was awarded the Patron's Medal of the Royal Geographical Society. He has been the President of the Alpine Climbing Group, Vice-President of the BMC, and is most proud of being the President of the Alpine Club and receiving the Piolet d'Or Lifetime Achievement Award in 2011, following on from Walter Bonatti and Reinhold Messner. He has published several notable books, including *Big Wall Climbing*, *Shishapangma Expedition* (co-written with Alex MacIntyre), *Up and About* and *The Ogre: A Biography of a Mountain.*

Russell Staiff is an adjunct fellow in the Critical Heritage and Tourism programme at Western Sydney University, Australia, and an adjunct professorial fellow in the Architectural Heritage and Tourism programme at Silpakorn University, Thailand. His research interests are twofold: the various intersections between cultural heritage, communities and tourism, with a particular emphasis on Southeast Asia, and heritage as a sociocultural phenomenon. He has recently co-edited the volume *Heritage and Tourism* (with Robyn Bushell and Steve Watson) and is the author of *Re-imagining Heritage Interpretation: Enchanting the Past/Future.*

Emma Waterton is Associate Professor in the Geographies of Heritage at Western Sydney University, Australia. She was a Research Councils UK (RCUK) academic fellow at Keele University from 2006–2010 and a DECRA fellow at Western Sydney University from 2012–2016. Her research explores the interface between heritage, identity, memory and affect in both Australian and international contexts. She has published nineteen books, including the monographs *Heritage, Communities and Archaeology* (co-authored with Laurajane Smith; 2009), *Politics, Policy and the Discourses of Heritage in Britain* (2010) and *The Semiotics of Heritage Tourism* (co-authored with Steve Watson; 2014).

Steve Watson was Professor of Cultural Heritage at York St John University, UK, where he taught a range of subjects in the cultural sphere. He made a substantial contribution to the field of Heritage Studies through his active development of theory that explores the relationship between representational practices and the performative encounters and engagements of tourists with heritage places. His most recent publications include *The Semiotics of Heritage Tourism* (co-authored with Emma Waterton), *The Palgrave Handbook of Contemporary Heritage Research* (co-edited with Emma Waterton) and *Heritage, Affect and Emotion: Politics, Practices and Infrastructures* (co-edited with Divya Tolia-Kelly and Emma Waterton). Sadly, Steve passed away in 2016, before the completion of this volume.

Ross J. Wilson is Professor of Modern History and Public Heritage at the University of Chichester, UK. He has written widely on issues of museum, media and heritage representations. This work has been published in *Representing Enslavement and Abolition in Museums* (2011), *Landscapes of the Western Front* (2012), *Cultural Heritage of the Great War in Britain* (2013), *New York in the Great War: Making an American City* (2014), *The Language of the Past* (2016), *Natural History: Heritage, Place and Politics* (2017) and *Gender and Heritage* (2018).

Acknowledgments

This collection owes a debt of gratitude to the late Professor Steve Watson for steering its inception with his indelible imagination and creativity. Many of the key themes of the volume were formulated over cups of tea and biscuits in Steve's cosy kitchen, while Beryl the dog ate carrots on the mat. Sincere thanks to Mrs Watson for generously providing photographs of Steve, and for your many other kindnesses. We are also extremely grateful to the publishers, Routledge (in particular Faye Leerink and Ruth Anderson), and the series editors (Emma Waterton and Divya P. Tolia-Kelly) for their flexibility and patience as we navigated our way through the painful setback of Steve's passing.

Many thanks are due to the contributors to this volume, all of whom have dedicated considerable time to their manuscripts – drafting, honing and polishing what we consider to be works of theoretical acuity and, in many cases, breathtaking storytelling. We hope the breadth of approaches parcelled between these pages – from critical engagement to narrative to photographic reflection – captures the eclectic complexity of what is a surprisingly unresearched area: affect at the frontier. As one might hope from a volume about 'frontiers', some of our authors draw on controversial and/or personal accounts in order to flesh out and explore some emotionally potent topics. Such vulnerability is no small feat, so we are also grateful for their brave writing!

We, alone, take credit for any errors or omissions.

Hayley Saul and Emma Waterton, March 2018

Introduction

Rethinking frontiers: transformation, exploration and adventure

Hayley Saul and Emma Waterton

'Glorious Coronation Day news! Everest – Everest the unconquerable, has been conquered. And conquered by men of British blood and breed', read the headline of the *News Chronicle* on 2 June 1953, four days after Edmund Hillary and Tenzing Norgay reached the summit. Notwithstanding the clear attempts to link the moment to the 'dawning' of the second Elizabethan Age, the newspaper's reaction epitomizes a peculiar set of values and practices that have come to encapsulate exploration's genealogy: heroism, endurance, optimism, courage and the quest for glory. In this instance, Everest was conquered in an act of aggression and a show of British masculinity, with the heroes of the piece rendered exceptional by both biology and heritage (which sits against the striking absence of any reference to Norgay). As Kennedy (2014) notes, the term 'exploration' first entered popular usage in the mid- to late eighteenth century, with 'explorer' coined later in the nineteenth century. The emergence of the new prestigious term of 'explorer' materialized more or less alongside the mythologizing of the 'frontier' in the work of Frederick Jackson Turner and his 'Frontier Thesis' (Turner 1893), which documented the role played by the westward expansion of land-grabs in American history, and the concomitant process of Americanization. Perhaps the most enduring image to emerge from Turner's thesis – and echoed in the *New Chronicle*'s headline – is that of a hostile and seemingly unconquerable wilderness, at once mighty and powerful, being transformed and settled, little by little, until it receded – along with the influence of Europe. It is the image of the American frontier giving way to the 'land of opportunity'.

It is this enduring conceptualization of the 'frontier' that came to occupy the centre of so many projects of conquest that were to follow, where the term acted as a 'metaphor for national development in its material and ideological senses' (Fold and Hirsch 2009: 95). While, in reality, these frontiers often were, and at times continue to be, excruciating spaces of assimilation, violence and exploitation, what has lingered – in both popular and scholarly discourse – is a notion of the frontier as that which lies at the very limits of human experience, where it effortlessly bleeds into the realms of mythology. There, 'empty', stubborn and inhospitable spaces are transformed into places of abundance by virtue of the strength, inquisitiveness, courage and boldness of those who seek not just to survive them but to control them.

The retelling of some of the earliest voyages has left similarly heavy mythological legacies. By way of example, Patagonia ('Patagón'), popularly translated as the 'land of the big feet', is a reference to an Indigenous population of giants first described in the journals of Ferdinand Magellan (Nesvet 2014). Likewise, until the eighteenth century, when recreational mountaineering took hold, popular opinion held that the European Alps were the abode of dragons (Fleming 2002). Though the details of its mythology vary between regions, in numerous accounts – from Bering, Cook, Livingstone and Amundsen to Falcon Scott and Shackleton – the frontier is sanctified as a wrathful foe in both the European and American imaginary. Pitted against this monolithic and primeval wilderness, or landscape of risk, the explorers and early settlers in such tales are elevated to the apotheosis of humanity. Commonly, the two critical, affective categories that perpetuate this frontier imaginary relate to 'endurance' and 'triumph', virtues that are commonly attached to Western, male figures. Other affective encounters, and other actors, are but backdrops or preludes to that moment of trial or victory. When Falcon Scott described how he 'knew that poor Oates was walking to his death' (Scott 1912) when he exited the Terra Nova Expedition tent, he made only cursory reference to the swirling blizzard of the landscape that he was entering and made no mention of the immensity of the certain, agonizing, cold death that awaited him. The nature of the experiential frontiers that the group faced were relegated to a side note, while 'it was the act of a brave man and an English gentleman' (Scott 1912) that captured his attention – those enduring values with which they entered Antarctica, and that he carried around like armour. Even as Scott's own cold-weakened fingers failed him during his last few hours of life, his writing clung to the qualities with which he faced the Antarctic frontiers, 'which has shown that Englishmen can endure hardship, help one another, and meet death with as great a fortitude as ever in the past' (Scott 1912).

Given the strong overtones of nationalism and identity formation resounding throughout these accounts (Larson 2011; Pettitt 2007), it is little wonder that popular narratives so easily entertained the perspective that 'the frontier makes the man [*sic*]', bolstering a mythology that remains inherently uneven, racist, exploitative and chauvinistic. Less easy to understand is that these mythologies, and particularly the concept of frontier that sustains them, have thus far escaped sustained academic scrutiny – certainly beyond the discipline of history, in any case. Instead, while we have seen the growth of scholarly interest in borders, boundaries, lines and zones (Newman 2003, 2006; Silberman *et al.* 2012; Van Houtum 2005; Wilson 2017), there is a curious dearth of academic engagement on the nature and meaning of 'frontier'. And so, the myth continues. It is resurrected time and again, and is frequently rehearsed in today's media, where casual references to 'internet frontiers', 'digital frontiers', the 'western frontier', 'the frontier of youth', 'economic frontiers', 'space frontiers', 'resource frontiers' and 'Earth's final frontier' abound.

With this volume, we propose a re-examination of 'frontiers', and seek to furnish the term with a more nuanced understanding. In large part, this new kind of understanding emerges directly from the work of our authors, but we want to use this Introduction to lay out our own stall, so to speak, and set down an overarching consideration of human–frontier relations, which we see as being temporal, mutually constitutive and transformative. As such, we commence the collection with a redress of the *a priori* rendering of 'the frontier' as a wilderness backdrop against which humans are pitted. Instead, we ask what other frontiers we might encounter, and in which 'spheres' of our lives? In asking these questions, we seek to recognize that, in experiences of a range of landscapes, in risky physical and social endeavours, and across the full complement of human experience, a more complex and subtle emotional geography may be charted – something that has evaded scholarly debate thus far. For example, reading past the dispassionate discourses of endurance and calculated risk, Falcon Scott's own journal betrays entanglements with frontiers in both a physical and social sense, particularly as his fortunes decline and he opens himself up to impending disaster. In his mounting uncertainty on the return journey, and with the dangerous Beardmore Glacier with its yawning crevasses still to traverse, he writes: 'Well, it is something to have got here, and the wind may be our friend tomorrow.' Though he invites the icy landscape into a bond of friendship in these final days, he also inflects this bond in his writing with a submissiveness that identifies the frontier more firmly with 'the will of Providence'. Thus, in the course of his journals, dominant expressions of a courageous resistance to 'an awful place' (Scott 1912), which at first refused to engage with affective registers beyond those endorsed as part of the heroic narrative, gradually give way to subtle and complex glimpses of a far more nuanced set of frontiers – physical, social, psychological, emotional.

It is this full swathe of affective encounters at the peripheries of human experience that we seek to explore in this collection of essays, and in doing so investigate the mutually constitutive relationships between humans and their more-than-human co-actors at/on frontiers, whatever their guise. With this in mind, the volume presents a series of experiences of frontiers as spaces of transformation, exploration and adventure. It reflects, principally, on the nature, form and experience of frontiers in terms of individual engagements, cultural encounters, extreme or challenging experiences, danger and risk, as well as the idea of frontiers as loci of transformation and transgression. Our interest is in frontiers as experience, as encounter, as place and as metaphor, and in each case constituted by affective and emotional registers. Contributions have been gathered from scholars, explorers and writers who collectively offer accounts of their own experiences that are reflective and/or scholarly, as well as those who use primary source material to construct insightful interpretations from geographical, historical, feminist, post-colonial, anthropological, cosmopolitan and cultural studies perspectives.

Rethinking frontiers

Although there is a long history of scholarship revolving around the concept of 'the frontier', particularly that which explores it as a key motif in nation-founding myths, surprisingly little attention has been devoted to continually and critically rethinking the *meaning* of the term. It is a concept, it would seem, that has fallen on hard times and almost drifted from academic scrutiny (but see articles in the special issue of *World Development* entitled 'Reconfiguring Frontier Spaces': Rasmussen and Lund 2018; Watts 2018). Yet, it is still used, and frequently; and with it come images of conquest, boundaries, dominance, the periphery and contact zones. Our first scholarly encounters with the idea of 'the frontier' can be traced to our undergraduate years, and to the seminar rooms of Archaeology departments in both the United Kingdom and Australia. There, we came across the controlled frontiers of the Roman Empire, which stretched across Europe and were punctuated by numerous military sites and defences, many of which were memorialized by their inscription onto the World Heritage List in 1987. Famous amongst these sites is the north-western monumental complex of Hadrian's Wall in Britain, which is imagined as a 'meeting place between people of different cultures and with different interests' (Collins 2012: 5). In addition to its prominent defensive role within the Roman Empire, the wall has been of interest for the role it has played, geographically, as a physical border between England and Scotland (Hingley 2010). In this discourse, it has emerged as a key borderline encircling British imperial space, and thus is regularly imagined as a 'spatial and physical signifier for a definition of the boundary of English identity' (Hingley 2010: 228; Witcher *et al.* 2010). As Witcher *et al.* (2010) note, however, it is studies of Hadrian's Wall as a 'Roman frontier' that tend to occupy the foreground of academic inquiry, with military precision, planning and the construction that went into its building and use dominating the narrative at the expense of other interpretations. During those same undergraduate years, we were also introduced to Frederick Jackson Turner's 'Frontier Thesis' and the classic of all frontiers: North American settlement.

While we are inevitably influenced by our early scholarly engagements with the physical frontiers of the Roman Empire and North American settlement, as Watts (2018: 480) has argued, 'land is only one part of the story'. Our interest in this volume, then, lies with newly emerging configurations of frontiers as social and emotional encounters, and here we borrow from the work of Rasmussen and Lund (2018: 388; original emphasis), who argue that the frontier should no longer be construed as empty geographical space; rather, it 'is not space itself. It is something that happens *in* and *to* space. Frontiers *take* place. Literally.' Thus, while we continue to include a sense of contact or friction in our conceptualizations of 'the frontier', we, like Watts (2018), prefer to adopt a deliberately abstract understanding of the term that is only loosely associated with 'limits' and instead engages with what Anna Tsing (1994: 279) has labelled the 'zones of unpredictability'. This is not to suggest that we are interested in parring the term back so much that it becomes an empty referent. To the contrary, we want to push things a little further in order to understand frontiers as powerful and extraordinary social *encounters* that are imaginative, affective and transformative; and we appreciate that they are encountered by ordinary people in moments when such people attempt to negotiate difference. Frontiers are made, unmade and remade in these moments of encounter.

To give further shape to the notion of frontier that lies at the heart of this book, the following sections unpack the three bodies of thought that have most helped us develop our understanding.

Affective frontiers

The first theoretical logic that we see as key to this volume is an interest in theories of affect, and the concomitant argument that 'worlds are sensed, not just seen' (Greenhough 2010: 43). This sort of logic requires us to rethink and reposition the body and its capacities to think, react, feel and sense in our academic work. As a starting point for this line of thought, we draw on the developing seam of critique that has emerged recently in Heritage Studies that troubles the centrality of representational thinking (see Waterton and Watson 2014; Tolia-Kelly *et al.* 2017), and that derives an understanding of affect from those modes of critique that revolve around the more-than-representational, across which we can see the clear foregrounding of notions of practice and experience. There are precedents to this school of thought, of course, which emerge from the cognate fields of Cultural Geography, Social Psychology and Cultural Studies. Scholars such as Judith Butler (1993, 1997) and Karen Barad (2007) have long been recognized for their attempts to decentralize discourse and textuality as the 'main business' through which academics think about their data, and their work has guided many attempts to open up intra-action processes. Of these, we are particularly intrigued by the work of Barad, as our interest lies with the ways in which data and events happen to intersect and influence our *selves* and our *lives*, and the ways in which the agency of humans (other persons, interviewees and so forth) and non-human matter (places and objects) influence and shape our thinking and our writing in ways that co-transform each other.

In taking up a more-than-representational flavour, we are signalling our interest in rethinking the limits of the body by bringing into focus the centrality of feelings, emotions and affect, and reconceptualizing aspects of experience and engagement in ways that are linked with the senses. In so doing, we take up Gorton's (2007: 342; original emphasis) argument that 'our feelings are not just registered in our conscious awareness but are *felt* and enacted by our bodies'. Some of the most essential characteristics of affect, as it is described in the more-than-representational literature, are its dislocation from the conscious subject, its unconscious transmissibility between bodies, and the ways in which it is implicated in 'atmospheres'. In her introduction to *The Transmission of Affect*, Teresa Brennan (2004: 1) famously asked: 'Is there anyone who has not, at least once, walked into a room and "felt the atmosphere"?' Affect, for scholars such as Brennan, is thus enacted in embodied physiological, psychological and neurological registers. Intensities, moods and atmospheres are transmitted into an individual body, where they change the body's biochemistry, if only for an instant, so something that was not there before is now *there*. Brennan also speaks of the origins of atmosphere and affect in social and psychological terms. This is important because it broadens the context of affect from purely individual sensation, and once it admits the social into its nexus it allows for the consideration of social practices such as representation and language. In Brennan's concept of affect, we thus have the foundations for a more-than-representational domain of human and non-human action, where meaning is constituted in diverse and richly textured sensory worlds. The problem of affect as a *something* that is trapped beneath a layer of subjective expressivity is addressed, therefore, by seeking its transmission in the sub-expressive layer of the social world rather than in the physiological and neurological systems of the individual body.

Concepts such as affect, mood, atmosphere and encounter are deployed by several of our contributors in order to rethink, and re-enact, the figure of 'the frontier'. As those contributors point out – sometimes explicitly and sometimes implicitly – such a move requires us to acknowledge that frontiers, as spaces of encounter, bring us into contact, physically and/or imaginatively, with people, places and experiences that are shot through with affective registers of fear, boredom, apprehension, joy and pain, to name just a few. They taste, feel, sound, look and smell, all of which impact on the body and its experiences of frontiers. How we begin to discover and describe such frontiers, when they are shaped and expressed in moments and spaces that are simultaneously intensely personal, intimately shared and ultimately social, is a central focus for many of our authors, whose vivid descriptions and analytical work point to their sensual, emotive and embodied aspects. Frontiers, in these cases, are momentarily, and in some case permanently, registered on and by the body.

Imaginative frontiers

Our second theoretical logic takes us far from the idea of the frontier as solely geographic space and into the realm of the imaginary. Amongst theories of 'the imaginary', we suggest that there are four particularly influential voices that have positioned the concept within an individual–community–society framework: Jacques Lacan, Cornelius Castoriadis, Benedict Anderson and Charles Taylor. While, by charting their individual work, we have the benefit of a rough chronology, the outcome of such a linear framework is that each perpetuates a core–periphery tension in which 'the imaginary' sits uneasily. Instead, we find their overlaps rather more significant than their sequencing. For Castoriadis (1987), the social imaginary is people's scope for imagining plus the actual outcome of that imagining. In other words, the imaginary describes the shared abilities of people and groups to conceive realities beyond the strictures of social norms and then act upon those conceptions. His emphasis is on imaginative unity (Strauss 2006), rather than heterogeneous imaginaries in a state of friction. As such, the imaginary proceeds outwards from a unified core of individuals, intoning with a creative progress: imagination is the seed of creativity that drives progressive development.

Conversely, Lacan's (1977) imaginary assumes more of a hyper-individual aspect. It is a fantasy: that is, a personalized interplay of the ego and the images to which it is exposed. Within such a framework, society remains 'real', an enveloping 'cloud' of norms in contrast to individual fantasies of it. Situated almost as a membrane between the individual and society, the imaginary for Lacan (1977) is an interpretive threshold by which society is received by the individual, and because of this the imaginary is always bounded by the scope of society itself. Anderson (1983) and Taylor (2002) share a similar understanding of the imaginary as indivisible from society. The fluid reality of what society *is* is negotiated in an abstract space of imagination that is mediated by materialities, images and other representational forms (Strauss 2006). This means that the imaginary (and therefore society) is highly fluid and amorphous, limited only by the creativity of actants and the communicative dynamics of their cohesions. Taylor (2002) diverges at this point to include an extra layer in the relationship, positing that group practices that propagate the imaginary undergo, in the first instance, a theorizing stage that then informs the practices by which the imaginary comes into being.

All of these theories afford materialities and representational devices supremacy in the types of social imaginaries that emerge. This, by extension, suggests that the breadth and colour of our imaginaries are bounded by our limited abilities to recycle the mediated prompts around us creatively. It is a wholly material and representational affair. But what of the imaginary's *frontier* – the less well-theorized affective horizon where materialities and images lose their grip, and where there may be no experiential precedents? This is the space of future becomings. It is charged with a precognitive sense of 'the possible', and sensations that activate the inspirational. We see glimpses of this realm in the narratives of many of our contributors. Though we might expect this sort of 'questing' to be intimately linked to a progressive ideal, a pursuit of truth, many of our authors shun such a deformation of the fleeting, ephemeral sensations that are gifted. The muffle of silence, the contemplation of solitude, some snatched seconds of elation on a summit, a homesick look over the shoulder that simply brings appreciation for somewhere else – these are humbler truths, not marks of progress. For many of our authors, the imaginary is disrupted at the frontier, replaced by a quest for authentic sensation. Far from referring to some sort of Romantic authenticity – one that relates to an origin that is deemed to be more genuine than what comes after – our authors pursue their authentic selves at the imaginary's frontier by reflecting on the repatterning of novelty that these experiences bring.

Anthropogenic frontiers

It seems fitting, and a reflection of current environmental anxieties, that this inspection of the nature and meaning of frontiers should draw forth, in its third and final theoretical logic, strong intimations of the Anthropocene. Here, we follow the prompts of Cronon (1996) and Lorimer (2015), who ask us to consider *what comes after* a Nature–culture frontier. That is, in recognizing that wilderness, too, is a human creation, how do we interact with landscapes and environments in ways that do not romanticize a Nature that is somehow divorced from human concerns and therefore susceptible to control by them? If the wild, or wilderness, is not some sort of pristine Nature or a place of redemption from corrupting civilization, what *is* the wild frontier? For Cronon (1996), wilderness is a mirror of ourselves, our own latent desires and cravings. To recognize our changing selves, we need look only to historically evolving concepts of wilderness that have underpinned virtually every environmental management decision since at least the eighteenth century. There, and then, the term had connotations of savagery and barrenness; a dangerous liminal zone where the edges of civilized morality drew a roadmap straight to Satan (Cronon 1996). It was something to be avoided at all costs. By the nineteenth century, however, the term had undergone something of a remarkable and swift U-turn, at which point the very same unholy terror of the wilderness took on the more generic, though powerful, visage of the *sublime*. As a result, wilderness was no longer the sole purview of hellish creatures, but a space for humans to pit themselves and, in proving their moral mettle, realize the divine (Cronon 1996). Ideas of a terrifying primordial wasteland gave way to the wilderness as a sacred temple. Suddenly, striking out into the 'wild' delivered prestige to the frontier adventurer, and wilderness's value started to be appropriated by elite tourists seeking to play out Romantic ideals (Cronon 1996). Such a stark reimagining of the wilderness frontier between the eighteenth and nineteenth centuries demonstrates the human 'taint' on something envisaged as pristinely Other than us.

Instead, the human–Nature divide squeezes and folds, diffusing and reciprocating in inherently unstable ways. Tsing (2003) finds it impossible, in light of this, to locate frontiers at an edge. Instead, she envisages them as 'projects in making geographical and temporal experiences' (Tsing 2003: 5100), and such inflections of *action* are evident throughout this volume. The ethical imperative of such nature–culture meditations problematizes an ideology of the frontier where wilderness is closer to divinity because it has led our 'civilized' species to the delusion that it is our *real* home, to the extent that,

Cronon (1996: 17) notes, 'we give ourselves permission to evade responsibility for the lives we actually lead'. With such an insistence that we retune our eco-ethics, the Anthropocene is, arguably, the most important frontier in which we are currently entangled. Throughout this volume, our contributors tease apart not just what it *feels* like to confront new and rehearsed wilderness frontiers, but how to harness those emotions in order to manoeuvre inquisitively through other frontier spaces.

The collection's structure

As we outlined above, *Affective Geographies of Transformation, Exploration and Adventure* offers new insight into the nature and meaning of frontiers. Although these thematic concerns – affect, imagination and the Anthropocene – are present across all twelve contributions, we have adopted an alternative set of concepts around which to organize the volume: transformation; exploration; and adventure. These, it seems to us, are sufficiently robust to provide the collection with a relatively straightforward 'skeleton' but will also allow the important complexities and interconnections identified above to permeate the book in its entirety, rather than attempting to absorb each of them into a single section. Thus, while these organizing concepts will no doubt seem entirely unsurprising when read against the history of academic interest in frontiers, we have deliberately invoked them here to unsettle them and take account of the ways our contributors and their scholarship have remade and reused them.

Part I, 'Transformation', commences with a contribution by Russell Staiff (Chapter 1), who is at once both extraordinarily personal and erudite in his examination of his delicate negotiations of race, sexuality and desire as a young man in the 1960s. Focusing on the oppositions of borders and frontiers, his chapter tells the tale of private discovery, prompted by a chance encounter with James Baldwin's *Another Country* (Baldwin 1965) in 1966. Staiff's story is deeply affecting – filled as it is with moments of wonder and shame and the porous edges between 'normality' and fantasy – and unspools around a journey that transgresses both sexual and racial frontiers. His contribution culminates in a critical appraisal of a range of frontiers – social, cultural, political and historical – and details how, in working through and across his own sexual frontier and his negotiations of difference, he unwittingly reinforced racial exoticism and boundedness.

Chapter 2 pushes the volume a little further back in time, to the early to mid-nineteenth century, and examines the published work of women travel writers and their experiences of Spain. Three frontiers interest Steve Watson and Emma Waterton in this chapter: the geographical frontier of Spain; the historical-affective experiences of female travel writers; and a personal frontier. The latter comes about because of the untimely passing of Steve part-way through his research on the 'Spanish Imaginary', making this chapter, itself, a frontier of loss but also of celebration and love. The authors discover through their bricolage of notes and gathered thoughts that the roots of affect are not just to be located in ourselves, but in our surroundings, which generate an 'inter-subjective space' of emotional layers. This social realm is where meaning is constituted in a sensory dimension saturated with shared affect.

Denis Byrne continues with the theme of frontiers of loss in Chapter 3 through his reflections on his investigations into the landscape signature of the 1963 Gunung Agung volcanic eruption in Bali. Constructed across at least three temporal dimensions – the eruption itself, his 1994 landscape archaeology, and his contemporary reflective contextualization – the chapter evokes a sensation of time compressing, extending, flowing and punctuating. We are left with the beautifully crafted impression that the temporality of the frontier is abrupt only as a factor of the scale in which it is experienced. The chapter concludes with a reflection on the temporality of decay in light of the inevitability of transformation.

Retaining the theme of deep temporalities set by Watson and Waterton and Byrne, in Chapter 4 Ross Wilson uses dinosaur parks to explore humanity's place within a world that has witnessed multiple, extended extinctions. We are led through a historical appreciation for the changing ways that dinosaur replicas have shaped the human–Nature relationship. In particular, popular tropes in nineteenth-century dinosaur parks that positioned extinction events as 'things of the past' and dinosaurs as maladapted creatures set the stage for greater dislocations of humans from the concerns of their more-than-human antecedents. But, as Wilson notes, in our current Anthropocenic era of extinction threat, these same parks have become places to rethink our coexistence with earth-others, rather than pursue modernity's hard-line project of distancing and alienation.

Part II, 'Exploration', brings together stories of and by some of the world's most renowned explorers. Through these often first-hand accounts of physical, emotional and sometimes spiritual frontiers, these contributors offer unprecedented insights into their personal confrontations with affectively charged landscapes. In Chapter 5, Doug Scott, CBE, best known for his role in the mountaineering team that made the first ascent of Everest from the south-west face, relates the first ascent of the Ogre in the Karakoram, one of the most technically difficult climbs in the world. In the vein of a classic survival story, Scott recounts how fleeting moments of calm contemplation upon the summit quickly gave way to grim prospects for their safe return after Scott slipped and broke both legs at the ankles. The captivating narrative betrays the meticulous attention to detail that made Scott one of the best climbers in the world; we experience each crucial foothold and equipment placement as the chapter systematically builds towards its frontier. But it is not a physical frontier to which Scott ascends in his writing; indeed, barely a moment is dedicated to the triumph of the summit. What Scott is interested in is the frontier of survival, and how that edge between life and death is entirely intertwined with the care of others. Though his survival was a profound demonstration of grim determination on his own part, it is also in the tenacious work of his climbing partners that Scott locates the nature of the frontier, with their perseverance restraining the risky edges of the endeavour from overwhelming the team.

In Chapter 6, Robin Hanbury-Tenison, OBE – an explorer who, among many expeditions, is renowned for being the first person to cross South America overland at its widest point – relates his simple ambition to be alone in the Sahara. Though initially this ambition might seem the antithesis of Doug Scott's frontier as characterized by unique bonds of camaraderie, we come to learn from Hanbury-Tenison's beautiful tale that even a place that might feel desolate is irrigated to life by shared memories. Hanbury-Tenison's Tuareg companions reveal a rich, animated desert landscape of rock art and nomadic culture. The chapter viscerally evokes how this desert landscape sits at the frontier interface of silent emptiness and teeming life.

In Chapter 7, Hayley Saul uses the Himalayan explorations of early ethnographer and Tibetologist Alexandra David-Neel to explore the mortal frontier. As one of the earliest explorers to enter the 'forbidden kingdom' of Tibet, David-Neel's travel diaries represent a substantial contribution to the documentation of popular religion at the turn of the twentieth century. These dark stories narrate unusual encounters with a malleable mortal frontier where the boundaries of life, death and transformation are eternally redrawn. Mediating this frontier are a range of materialities that act to simulate death-like experiences, allowing practitioners of numinous rituals to experience death-states at an emotional level in order better to equip themselves spiritually. Saul explores how this expanded notion of death can resource debates in a contemporary era of extinction.

Part II closes with a contribution from mountaineer Tim Macartney-Snape (Chapter 8), who is renowned as the first Australian to climb Everest and the only person to ascend it from 'sea to summit'. Macartney-Snape recounts his experiences on two expeditions: Annapurna II and Everest. His reflections on these peaks explore the shift from 'unknown' to 'familiar' as something akin to befriending risk. Whilst Macartney-Snape centres the risky 'wilderness' as the place where an ethic of curiosity arises, and a source of worthy endeavour, the wilderness that he knows is not 'Other'. Instead, his expeditions can be likened, in many ways, to a form of socializing with the mountains: he is enriched by the slopes as they unfold before him and seem to urge him to familiarity. He views the mountain's panoramas as if through shared eyes, eager to know the mountain and be known to it.

Part III, 'Adventure', begins with Felicity Picken's haunting exploration of the altered states of being that are induced by the narcosis that may occur during deep-sea free diving (Chapter 9). Here, Picken challenges us to reconsider what happens when we transgress our landlocked being and experience unexpected and creative transformative dimensions that occur in the world's least explored region: the deep blue. Using divers' testimonies and the short film *Narcose* (2013), Picken probes the depths of the affective, physical and cognitive states brought on as the diver descends, the pressure increases and the nitrogen seeps into the body's tissues.

In Chapter 10, Jenny Hall picks up on this risky frontier and explores the role of risk as a type of play in the Scottish Highlands. Her focus is particularly on how women encounter mountaineering and negotiate the affective possibilities of 'the unknown', when the realm of risk is otherwise a highly masculinized environment. In Hall's analysis, fear comes to act as a 'fail-safe' valve that prevents risky play from transgressing the frontier into unpredictable danger. In acknowledging her fear whilst mountaineering in the Cairngorms during a day of inclement weather, she directly challenges the 'masculinized leadership' style of her female companion, who seeks to acquire greater symbolic capital through competitive endeavour at Hall's expense. In doing so, Hall sheds the contamination of masculinized 'ways of mountaineering' and plots her own novel relationship to the frontier of risk.

Sticking with the sport of mountaineering, Esther Bott (Chapter 11) takes up the notion of risky landscapes as proving grounds, but reflects specifically on the way a claim to 'wilderness' locates the climbing traveller as something authentic and different from the tourist. Bott transports us to the remote rocky outcrop of Li Ming in southern China and inspects the mediatized production of the climbing frontier imaginary at this secluded crag. She discovers that three imaginaries are at play for climbers drawn to Li Ming: the tourist–traveller seeking a cultural exotic as a marker of authenticity; the adventurer seeking themselves by confronting their own notion of wilderness; and the climber who searches for the frontier of ability and worthiness against the rock-face.

Drawing the volume to a close in Chapter 12, Rosemary Kerr locates the formation of Australian nationhood within a pre-industrial frontier, a supposedly authentic rural landscape that is not positioned as 'wilderness' but instead as a forge where wilderness gives way to industry. She explores 'the road' as the enduring contemporary connection to frontier legends that have been shaping Australian identity since the drovers, shearers and bushrangers expanded the sphere of workable land in the nineteenth century. These figures, Kerr claims, epitomize the resilience, endurance and tenacity of an Australian national spirit. Such frontiersmen qualities went on to shape the Anzac legend during the First World War. Kerr's chapter explores the meanings of 'the road' and the locations of a changing frontier that have shaped the cultural consciousness of the Australian nation.

References

Anderson, B. (1983) *Imagined Communities: Reflections on the Origin and Spread of Nationalism*. London: Verso.

Baldwin, J. (1965) *Another Country*. London: Corgi Books.

Barad, K. (2007) *Meeting the Universe Halfway: Quantum Physics and the Entanglement of Matter and Meaning*. Durham, NC: Duke University Press.

Brennan, T. (2004) *The Transmission of Affect*. Ithaca, NY: Cornell University Press.

Butler, J. (1993) *Bodies that Matter: On the Discursive Limits of 'Sex'*. London: Routledge.

Butler, J. (1997) *Excitable Speech: A Politics of the Performative.* London: Routledge.
Castoriadis, C. (1987) *The Imaginary Institution of Society.* Cambridge, MA: MIT Press.
Collins, R. (2012) *Hadrian's Wall and the End of Empire: The Roman Frontier in the 4th and 5th Century.* London: Routledge.
Cronon, W. (1996) 'The Trouble with Wilderness: Or, Getting Back to the Wrong Nature', *Environmental History*, 1(1): 7–28.
Fleming, F. (2002) *Killing Dragons: The Conquest of the Alps.* New York: Atlantic Monthly Press.
Fold, N. and Hirsch, P. (2009) 'Re-thinking Frontiers in Southeast Asia', *Geographical Journal*, 175(2): 95–97.
Gorton, K. (2007) 'Theorizing Emotion and Affect: Feminist Engagements', *Feminist Theory*, 8(3): 333–348.
Greenhough, B. (2010) 'Vitalist Geographies: Life and the More-than-Human'. In: B. Anderson and P. Harrison (eds) *Taking Place: Non-representational Theories and Geography.* Farnham: Ashgate, pp. 37–54.
Hingley, R. (2010) 'Tales of the Frontier: Diaspora on Hadrian's Wall', *Journal of Roman Archaeology*, 78: 227–243.
Kennedy, D. (2014) 'Introduction: Reinterprating Exploration'. In: D. Kennedy (ed.) *Reinterpreting Exploration: The West in the World.* Oxford: Oxford University Press, pp. 1–18.
Lacan, J. (1977) 'The Function and Field of Speech and Language in Psychoanalysis'. In: J. Lacan (ed.) *Écrits: A Selection.* New York: W.W. Norton, pp. 30–113.
Larson, E. (2011) *An Empire of Ice: Scott, Shackleton, and the Heroic Age of Antarctic Science.* London: Yale University Press.
Lorimer, J. (2015) *Wildlife in the Anthropocene: Conservation after Nature.* Minneapolis and London: University of Minnesota Press.
Nesvet, R. (2014) 'Patagonian Giants, Frankenstein's Creature, and Contact Zone Catastrophe', *Essays in Romanticism*, 21(1). Online. Available at: http://online.liverpooluniversitypress.co.uk/doi/abs/10.3828/eir.2014.21.1.5. Accessed 12 March 2018.
Newman, D. (2003) 'On Borders and Power: A Theoretical Framework', *Journal of Borderlands Studies*, 18(1): 13–25.
Newman, D. (2006) 'Borders and Bordering: Towards an Interdisciplinary Dialogue', *European Journal of Social Theory*, 9(2): 171–186.
Pettitt, C. (2007) *Dr Livingstone, I Presume? Missionaries, Journalists, Explorers, and Empire.* London: Profile Books.
Rasmussen, M.B. and Lund, C. (2018) 'Reconfiguring Frontier Spaces: The Territorialization of Resource Control', *World Development*, 101: 388–399.
Scott, R.F. (1912) *Diaries of Robert Falcon Scott.* Online. Available at: www.bl.uk/turning-the-pages/?id=12878b6a-36b9-44db-a940-365b21bfe524&type=book. Accessed 26 March 2018.
Silberman, M., Till, K.E. and Ward, J. (2012) *Walls, Borders and Boundaries.* New York: Berghahn.
Strauss, C. (2006) 'The Imaginary', *Anthropological Theory*, 6(3): 322–344.
Taylor, C. (2002) 'Modern Social Imaginaries', *Public Culture*, 14(1): 91–124.
Tolia-Kelly, D.P., Waterton, E. and Watson, S. (2017) 'Introduction: Heritage, Affect and Emotion'. In: D.P. Tolia-Kelly, E. Waterton and S. Watson (eds) *Heritage, Affect and Emotion.* London: Routledge, pp. 1–11.
Tsing, A.L. (1994) 'From the Margins', *Cultural Anthropology*, 9(3): 279–297.

Tsing, A.L. (2003) 'Natural Resources and Capital Frontiers', *Economic and Political Weekly*, 38(48): 5100–5106.

Turner, F.J. (1893) 'The Significance of the Frontier in American History'. Online. Available at: http://nationalhumanitiescenter.org/pds/gilded/empire/text1/turner.pdf. Accessed 13 February 2018.

Van Houtum, H. (2005) 'The Geopolitics of Borders and Boundaries', *Geopolitics*, 10 (4): 672–679.

Waterton, E. and Watson, S. (2014) *The Semiotics of Heritage Tourism*. Bristol: Channel View Publications.

Watts, M.J. (2018) 'Frontiers: Authority, Precarity and Insurgency at the Edge of the State', *World Development*, 101: 477–488.

Wilson, H.F. (2017) 'On Geography and Encounter: Bodies, Borders and Difference', *Progress in Human Geography*, 41(4): 451–471.

Witcher, R.E., Tolia-Kelly, D.P. and Hingley, R. (2010) 'Archaeologies of Landscape: Excavating the Materialities of Hadrian's Wall', *Journal of Material Culture*, 15(1): 105–128.

Part I

Transformation

1 Taboo desires?

James Baldwin, African Americans, homoeroticism and the frontiers of mind/place/race and sex

Russell Staiff

Introduction: the year was 1966

With my cousin, four years older than me, I had taken the train from the country town where we lived in southern Australia into the provincial capital, Adelaide. It was one of those memorable events that break into the small world of seemingly bounded horizons of rural life. I do not remember what film we saw. After the movie, we arrived back at the railway station some time before our departure and found ourselves in a bookshop. I was looking for something to read on the long, two-hour journey home. My reading tastes, at the time, were dominated by comedy (Wodehouse and the like) and war thrillers (Alistair MacLean). All the MacLean titles on offer I had read, so I was perusing aimlessly. In a distracted manner I picked up James Baldwin's *Another Country* (Baldwin 1965). I can't recall what I thought this book may have been about, nor what piqued my initial interest, but when I turned it over and read the blurb, I was transfixed. *Homosexual.* The word tore through me like a storm. My heart raced. My legs felt weak. I looked around, sure that other customers had seen what I was looking at, had known that this book was taboo. My anxiety increased as my cousin signalled that we needed to leave. I hurriedly looked at the blurb again: 'a study of a cat's cradle of sex, and near relationships based on sex, chiefly between blacks and whites, and between homosexuals'. How I managed to purchase this book I can now only imagine. The embarrassment would have been acute. I had just celebrated my sixteenth birthday.

This essay charts a powerful triad of frontier/border crossings: the mobility of imagination, of places real and imagined and of sexual desires, obsessions and behaviours that criss-cross national boundaries, racial divides and notions of 'normality', that mobilize a transgressive and 'secret' self-made all the more potent by words like 'taboo'. My mode of analysis is remembrance, a type of archaeology of the self/body that attempts to recall the power of these frontier/border crossings and the implications of such a narrative for our understanding of the complex ways imagination, desire, corporeal travel, racial cartographies and sexual behaviour are animated and performed in motion, where everything is improvised, fluid and lacking fixity (see Eguchi

2015). Only in retrospection is there a type of stasis, a pinning down in analysis, the illusion of solidity when, in fact, there was none at all.

I have chosen a narrative mode because of its constant drive forward, its relentless movement, its fusion of action and time, its ability to reshape time and place, its focus on character and its first-person capacities. But I also employ narration with extreme caution because of its often inbuilt teleology, its plot determination and its urge for resolution. These aspects of narrative will be eschewed. One thing that cannot be ignored is that my narrative/analysis is a representation of fluidities that are, almost wholly, experienced outside of representation. However, there is no way this conundrum can be easily circumnavigated.

How to describe a white teenager's obsession with the male African American body when living in rural southern Australia? Decades later, only the contours still exist. Maybe of importance was the train trip home from Adelaide, appropriately into the growing darkness of night, a darkening landscape streaming past the window outside, but cocooned in a dimly lit carriage compartment, where my internal gaze was on a very different landscape prompted by my reading. Certainly *Another Country* enabled so many things simultaneously. I was reading about something I thought was not possible to utter: love and sex between men. But it was somewhere else; the locus of this 'world' was not the farm where I lived and the town where I went to school. It was New York. And this sexual landscape, so earnestly desired and yet seemingly impossible ever to be realized except in ever more elaborate sexual fantasies, was, yes, in 'another country' and so far away. And yet, disregarding geographic distance altogether, this masturbatory space – of desire, fantasy and acute physical and bodily sensations of great need and exultant release – was inhabited by images of African Americans. These images were so readily fed to me by television, magazines and the cinema. I secretly began to collect pictures of African American singers, actors, athletes and those who made it into the print media (usually criminals). The folio of pictures and the novel (and I soon added *Giovanni's Room* (Baldwin 1963) to my collection) were hidden in a distant shed on the farm that stored sacks of charcoal used in burners in winter to ward off the frost in the growing season for hothouse vegetables. There, I could steal away from the oppressive 'normal' world of my everyday and into my forbidden world of teenage lust, where I imagined furtive meetings with African American men.

It was such a strange separation of the sexual self from the objects of desire; a constant moving between an imagined place (the USA and especially New York), people who had no physical presence in my daily life – African American men – and the intensity of masturbation. In the play *Equus*, Peter Shaffer (1977) ponders the mysterious way certain aspects of one's life are magnetized by sexual desire; the indeterminate mobilities of body, mind and the material creating in their coalescing what Freud termed a 'fetish'.

The following meditation is in two parts. I begin with Baldwin's novel of alienation, *Another Country*, in an attempt to re-enter the fictive world read on a rural farm in southern Australia in the 1960s. It combines memory work and an unexpectedly intense emotional response to rereading the novel in 2016. The second part is a partial and incomplete reflection on the issues raised by the frontiers of mind/place/race and sex in the 1960s and now. However, I begin with a note about 'borders/frontiers', a contextual and conceptual musing that informs the analysis.

Borders/frontiers: a Janus-faced phenomenon

The commonsense understanding of frontiers is that of space/place (an admixture of the imagined, the conceptual and the physical) lying 'beyond borders'. There is, persistently, the suggestion of something unknown, something tinctured with trepidation, wonder and desire. Critical to the frontier, in Western thought and action, is its hinged companion – border. The binarism is deeply cultural and social, political and geographical, historical and juridical, linguistic and economic, scientific and artistic. Henk van Houtum encapsulates the dynamism of this twinned phenomenology by assuming that borders/frontiers are complex spatial and social entities that are highly effervescent. He writes thus:

> A socially constructed border is a form and manifestation of self-repression. It suppresses the total potential of personal mobility and freedom by constructing a sphere of trust inside and a fear of what is out there, beyond the self-defined border. Yet at the same time, the world outside that is constructed by a border also expresses a desire, a wish, the longing to be somewhere else. It is the desire to experience and live the personal freedom despite or thanks to the fear of the unknown, the non-routine. That is the desire to turn to the other, the desire to cross the line. The unknown, the stories about the exotic and the mythical, the adventure, the wild or the culturally different, can work like the Siren song on our ears. A border therefore also reflects liberty, the desire to de-border oneself, to become stranger oneself. The desire to leave behind whatever is familiar, to close the door behind, to turn the key and [go] into the world … to become strange and stay strange.
>
> (van Houtum 2011: 58–59)

I'm attracted to the 'queer' flavour of this description, to the 'othering' that is seen as crucial to the idea of transgression, to delineation and to transformation, to attempted fixity and to the demarcation of inclusion and exclusion at the very moment when the liminal, the porous and the provisional exert a perceptual/conceptual ecology of constant and crucial cross-border processes and the idea of borderless realms.

In recent years, border studies have grown apace, especially after the collapse of the Soviet Union, the profound impact of globalization, the EU experience and the more recent assertion of regionalism/localism (Wastl-

Walter 2011). But following Foucault and other post-structural theorists, it is well understood that modernity and the construction of the self was via a proliferation of spaces and the multiple creation of distinctions and boundaries that produced 'others' (Foucault 1973; Gregory 2004). But such productions were in a time of radical critique of limits and so the frontier shifted from its nineteenth-century New Worlds conception (frontier lands) and its mid-twentieth-century scientific formulations (space, medicine) to a quest (individual and social) for a type of border defiance/infringement critique as 'beyond borders'. The current asylum/refugee predicament epitomizes the flux that constitutes any notion of 'frontier'.

Whether explicit or not, borders/frontiers are deeply embedded in most cultural logics; they are to a degree universal human experiences but expressed in specific cultural forms/norms. Language provides one clue. Every noun demarcates that which is not something else and in so doing defines a border and a space between this and that. Categories, in whatever epistemic field – from science to literature, art to geography, sociology to archaeology, literature to mathematics and so forth – are, on one level, an extension of the logic of nouns, the naming of things, and, on another level, abstractions that by their very nature have boundaries (that then gesture to that beyond the border line). A highly necessary and highly significant corollary of this is transgression, when phenomena migrate across borders, when borders are deemed permeable or when they fail or are removed, or when borders are limits that activate a desire for frontier phenomenon. Transgressions and cross-border thoughts/actions are as crucial as the borders themselves because, paradoxically, they make the borders so radiant. Frontiers, however, are not usually perceived as bounded; they are spaces of expansion, of possibilities, of utopias, of futures.

Time travel: the remembrance of reading James Baldwin's *Another Country*

What follows is a reconstruction of the past. The cliché of memory being like the shifting sands of the desert or a mirage is apt. In memory, linear time evaporates; time folds in on itself ceaselessly. Nothing is fixed. The chronology of events gets muddled. James Baldwin, the television series *Roots* (Margulies 1977), Robert Mapplethorpe's photography, Mark Twain's *Huckleberry Finn* (Twain 1884), Jimi Hendrix, Harper Lee's *To Kill a Mockingbird* (Lee 1960) are all in the mix but the chronology requires my concerted effort. Baldwin came before Lee; Twain after Baldwin and not long before Hendrix; *Roots* was still a decade away and Mapplethorpe's images of naked African American men later still.

So what to make of my remembrance of reading *Another Country*? The rereading, fully aware that I would be writing this chapter, occurred in New York in June 2016. The rush of memories was unexpected, but, interestingly, I did not readily relate the novel to my being in that city. The 'real' New York was, obviously, peripheral to the imagined New York of my teen years and the conjured metropolis, although now entirely lost to me, I fleetingly recall as a phantasm of immense imaginative richness. Nevertheless, the New York of my travels (I have been there several times) is organically linked to Baldwin's novel, to African Americans and to gay sexuality. While my professional interest was in the culture of New York (its museums, its theatre, its music, its dance, its architecture, its history), there was always the pull of being somehow immersed in an exciting subterranean world of gay sex and African American male bodies, of sensuality, of forbidden desires, of experiences where the racial 'other' had melded with my sexual fantasies. The eroticized black male body has been the subject of highly significant cultural/political analysis (see below) and in recent decades I have been acutely conscious of this critique – a critique that, ironically, has made New York all the more alluring because it has amplified the position of the African American in contemporary society, has turned a spotlight on race and racial/gender/sexual politics/divisions and, for me, has given a noble/heroic dimension to the ongoing struggle of African Americans for justice, equality, visibility and presence.

For a long time now, I have been trying to 'see' and 'feel' the United States through African American sensibilities. What began as a teenager's adulation of the black naked body has become a political/ideological position that extends to wherever race is utilized as a political and social tool. My support for Indigenous Australian land rights and sovereignty, and my years of anti-Apartheid activism against the white South African regime arose from a somatic process of desire/reflection/knowledge; of mental border crossing where a fetish had transformed itself into acute empathy.

While all this can be regarded as one of the end points of border transcendence, it is an intellectual response and not the somatic bodily response of a white sixteen-year-old living thousands of kilometres from Manhattan. In what follows, I want to focus on two passages in *Another Country*, both of which form part of Eric's recollections as he prepared to leave his lover Yves in France and return to New York. As he took a shower 'he was alone with his body and the water', and the thoughts of his life in New York before his 'escape' to Paris 'caused many painful and buried things to stir in him' (Baldwin 1965: 152). And what were those painful things?

> They were buried beneath the impossible language of the time, lived underground where nearly all of the time's true feeling spitefully and incessantly fermented. Precisely, therefore, to the extent that they were inexpressible, were these terrors mighty; precisely because they lived in the dark were their shapes obscene … he had nearly perished in the basement of his private life. Or, more precisely, his fantasies.
>
> (Baldwin 1965: 152)

Eric had grown up in a sterile emotional environment where his parents' world was one of 'habit' and 'courtesy', and where their separate and busy lives kept them away from home. Emotional sustenance came from the African American cook, Grace, and her husband, Henry, who managed the boiler system in the building's basement. Eric spent many hours with Henry but on one occasion he found him sitting alone in the furnace room:

> [W]hen he spoke Henry did not answer; and when he approached him, putting his hand on Henry's knee, the man's tears scalded the back of his hand. Eric no longer remembered the cause of Henry's tears, but he would never forget the wonder with which he then touched Henry's face, or what the shaking of Henry's body had caused him to feel. He had thrown himself into Henry's arms, almost sobbing himself, and yet somehow wise enough to hold his own tears back. He was filled with an unutterably painful rage against whatever it was that had hurt Henry. It was the first time he had felt a man's arms around him, the first time he had felt the chest and belly of a man; he had been ten or eleven years old. He had been terribly frightened, obscurely and profoundly frightened … He knew that what he felt was somehow wrong, and must be kept a secret; but he thought it was wrong because Henry was a grown man, and coloured, and he was a little boy, and white.
>
> (Baldwin 1965: 153–154)

Even after such a long time, the description of Eric and Henry is still affecting. I no longer remember the exact emotional terrain of my sixteen-year-old response, but what did surprise me was the unanticipated depth of feeling the erotically charged passage provoked in me in my recent rereading. At sixteen, I had never felt a man's arms around me, not even my father's. At sixteen, I was both frightened and excited about my same-sex attraction. Perhaps there was something deeply disturbing/comforting about substituting myself for Eric and in my imagination *feeling* the protecting, and yet somehow taboo, embrace of an African American man. Over there in Alabama – where Eric spent his childhood – something was *possible* that seemed impossible in a small, claustrophobic, rural town in Australia. I crossed, in my mind, a geographic, racial and sexual frontier as a way of coping with a life lived where borders were (seemingly) distinct and unassailable: rural towns, white, Anglo-Irish Australia, heterosexuality, religion, societal norms and gender roles, especially masculinity.

Indeed, the very next passage described vividly my own great fear – the fear of exposure – and the means of dealing with it, secrecy: 'But by this time he knew that everything he did was wrong in the eyes of his parents, and in the eyes of the world, and that, therefore, everything must be lived in secret.' Baldwin goes on to describe Eric's sensation of separateness produced by his feelings of anguish about the 'teeth of the world' and by his dreams/imagination; that while he may be outwardly 'normal', none of his friends had 'kissed Henry on his salty face'; what lived in his being was 'the brilliance and beauty and terror of desire' (Baldwin 1965: 154).

The second episode of epic consequence, after Eric's shower reverie about Henry, was his recollection of when the 'blow had inexorably fallen' (Baldwin 1965: 201). Baldwin describes Eric sitting by the window of the bungalow – with views of the Mediterranean – he and Yves had rented for the summer. A glass of wine and a cigarette in hand, Eric gazes out to sea, the sun sinking, the water 'on fire'. He recalls a hot and dusty Sunday afternoon, the town in Alabama listless and quiet. Eric is sixteen years old and walking on the edge of town with LeRoy, his seventeen-year-old friend. LeRoy! My memories were ablaze, my heart beat faster. Until his name appeared on the page, I had forgotten. Now, sitting in a New York café rereading the novel, I was swamped with a palpable recognition, with somatically charged memories: this was *the* episode. I was Eric. LeRoy was the object of my desire. The very name, LeRoy, still has such potency. And LeRoy was African American.

LeRoy worked as a porter in the local courthouse. He was 'tall and very black' and the two had been friends for a long time. It was a fraught friendship: people sniggered and whispered knowingly. On this particular hot Sunday afternoon, they are walking aimlessly and chatting about getting out of Alabama and heading north to one of the big cities. The fictional teens also imagine a world beyond the bounds of the rural town in which they live. Eric puts his arm around LeRoy's shoulder and LeRoy puts a hand on Eric's neck. A roadster of young folk – three white boys and three white girls – is upon them before they can jump apart. A great mocking laughter erupts from the car as it speeds past. They part instantly, Eric's face aflame. LeRoy chides him because 'You *ain't* supposed to be walking around this damn country road with no nigger' (Baldwin 1965: 158; original emphasis). They leave the road and withdraw to a clump of trees overlooking a stream. Eric declares his love for LeRoy and in their physical lovemaking they both experience the terror of the forbidden – not just male–male sex, that is bad enough, but the momentous implications of crossing the racial divide in a southern state. Enveloped in LeRoy's arms, Eric is alive to his body's sensations, to LeRoy's responses, his smell and the experience of 'anguished joy beneath the weight of his first lover' (Baldwin 1965: 160). The sex transforms Eric: 'what had always been hidden was to him, that day, revealed' (Baldwin 1965: 160).

It was the nature of the revelation that was so magnified in my imagination, then and now. This was the passage where I sexualized the African American male body. All those years ago, I had just read about two teenagers having sex – not just mutual masturbation between curious, hormone-charged boys, but a description that went much further, where feelings of love, attraction, yearning and passion were included. Until that moment, all these things had been visible to me only as heterosexual realities, in life, novels, film, television and the media. There, in Eric and LeRoy's story, was a kinship with others who shared the same male–male desires as myself, who shared the same secrets, the same hidden selves, and faced the same struggles and fears. This was the revelation and, for me, it was, for better or worse, unconsciously entwined with race.

A teen sexual urgency was thus fused with a powerful attraction to African American men. The fictive had clarified all sorts of emotional issues: that my own 'secret' sexuality seemingly could only ever be (safely) realized in a place far beyond the borders of my existential world, in New York, and in the orbit of black men. And so the far-off – in a frontier constructed as geographic, racial and sexual – was both a refuge and a never-to-be-realized goal that could nonetheless prove to be highly sexually arousing. The inner world seemed to be contained, safe and secure. It seemed to have borders within which I could escape and borders that kept the torments and condemnation at bay. I was never the subject of gay hate at school or physical abuse, but I was highly alert to the social ostracism of homosexuality in the language of the school yard: 'sissy', 'poofta', 'bum boy', 'faggot'.

LeRoy and his neighbourhood, however, also stood for something else, even back in 1966. Baldwin's writings conjured powerful scenarios of injustice: the poor 'black' end of town, the marginalized and the shunned. Did I see in the plight of Baldwin's African Americans and gay men my own alienation, my own minority status, my own marginalization? Maybe. For me, 1966 was not just the year of *Another Country*; it was also the year when I read Harper Lee's *To Kill a Mockingbird* and followed media reports of Cape Town's notorious District 6 evacuations. I think race and sex and the politics of social justice coalesced, but I'm sure I didn't understand the interconnections at the time.

Today, the intersections of race, politics and sexual desire still reverberate from the past and express themselves in somewhat sedate intellectual pursuits across diverse phenomena – African American art, jazz, African American cinema and television, African intellectuals, the 'Black Lives Matter' movement and the racial politics of the United States. The profound pain and suffering of the marginalized and the oppressed still too often turn on borders produced and policed by the racialized body, the ethnicized body, the gendered body, the sexualized body (and so on). The frontier world of 'the other' continues its powerful affect/effects.

The African American male body after deconstruction

Understandably, as a student of the visual arts, I could not resist the allure of the debates prompted by the photography of Robert Mapplethorpe (1983), and in particular his eroticized/aestheticized photographs of the African American male nude body. Just as seductive were the writings of black gay men, and, in this instance, Kobena Mercer's (1991) essay 'Skin Head Sex Thing: Racial Difference and the Homoerotic Imaginary'. This highly important and eloquent essay should be read in its entirety and I have no intention of trying to summarize it here. What is significant is that in the 1990s I was confronted by my own complicity in the objectification and stereotyping of the African American black male body as a white gay spectator; in the colonialist implications of my gaze (privileged, power laden, exploitative), and in the highly ambiguous position I had adopted with regards to race, sexuality and politics in my consumption of black culture and African American images of men (whether in 'high art' or in 'ebony' gay pornography).

It was not guilt that bubbled to the surface when I confronted the African American male body post-deconstruction; when I reflected deeply on my forays into the fantasies that depended upon a constructed racial and sexual frontier and where border transgression (and thus an assumed fixity of racial and sexual identities/borders) had been crucial to my masturbatory liberation from my existential sexuality-induced angst. Rather, what perturbed me was how I had justified the role African American men had played in my ability to cope with what Baldwin brilliantly described as the 'terror of desire'. My fetish for African American men had been a refuge, a space of solace. Somehow, New York, gay 'men of colour', all bordered entities, had been a psychic force that countered white hyper-masculinity, rural life, hetero-normative expectations and the fear of exposure. 'Gay African American men' acted collectively as an antidote to the white heterosexual men I encountered daily. Baldwin's *Another Country* was a gift: from that point on, my world was so much bigger and suddenly full of potential hitherto only vaguely imagined. The frontiers of race and sex gave my life a new, expanded shape of possibilities. Nevertheless, difference, in all its guises, means borders. How, then, to *think* about this after reading writers like Kobena Mercer?

Looking back, my fantasies seem to have been a type of (unconscious) resistance to the dominant and suffocating masculinist performances in a white, hetero-normative small rural community. Homi Bhabha's (2001) notion of 'vernacular cosmopolitanism' is taken up by Gregory Castle (2001) in his analysis of location in the colonial/post-colonial condition, where individuals at the local level, especially marginalized individuals, challenge, resist, ignore and coexist with 'universal' historical narratives, with the collectivized, with the hegemonic vectors within which they live. Reading Baldwin and, subsequently, idolizing gay African American men were, I think, a form of resistance.

But – and it is a big 'but' – I cannot ignore the ugly reality that there were indeed overtones of the 'noble savage' in my fantasies (as I had no physical contact with either New York or African Americans). Whatever the degree of exoticization, and however much my 'escape' ennobled the 'Other', such 'othering' was and is foundational to the colonial project and is riven with unequal power relationships. It produces stereotypes, both positive and negative. Colonialism, as Bhabha has shown, is a highly ambiguous space/place (Bhabha 1994). White fantasies about black bodies will always be problematic because, in the fantasies, those bodies are inevitably subjects and utterly dependent on racial difference. The end point, on rare occasions, might be empathy, but even this is achieved through a type of exploitation. Therefore, reading Baldwin as a white male teenager, despite everything, cannot be regarded as anything other than an act mired in ambivalence. It is fraught precisely because the racial, sexual, ethnic and gendered borders must remain intact if the fantasy is to be efficacious (see Williams 2004).

While Kobena Mercer confronted his own intellectual vacillations about being a black gay man looking (lustfully) at the erotically charged photographs of African American naked men taken by Robert Mapplethorpe, a decidedly white photographer, Essex Hemphill, in his introduction to the 1991 anthology *Brother to Brother: New Writings by Black Gay Men*, is far more clear cut:

> The post-Stonewall white gay community of the 1980s was not seriously concerned with the existence of black gay men except as sexual objects. In media and art the black male was given little representation except as a big black dick. This aspect of the white gay sensibility is strikingly revealed in the photographs of black males by the late Robert Mapplethorpe. Though his images might be technically and aesthetically well composed, his work *artistically* perpetuates racial stereotypes constructed around sexuality and desire. In many of the images … [t]he penis becomes *the* identity of the black male which is the classic racist stereotype re-created and presented as 'art' in the context of a gay vision.
>
> (Hemphill 1991: xxxix; original emphasis)

In 1972, James Baldwin, in his well-known 'rap on race' with the anthropologist Margaret Mead, said much the same thing:

> They come to you for the most part, as though you're some extraordinary phallic symbol … As if you're nothing but a walking phallus … no head, no arms, no nothing … actually the act of love becomes an act of murder in which you are also committing suicide.
>
> (Baldwin quoted in DeMarco 1983: 113)

Where does that leave a white teenager in rural Australia who is sexually aroused by images of naked African American men? It is not a simple analytical choice. The intellectual and conceptual response came many decades later. What is still a source of wonder is the *emotional* and *somatic* response to reading Baldwin, the image of Eric touching George's tear-stained face, the searing remembrance of Eric and LeRoy making love. And, above all, a retrospective understanding of what those descriptions, read in a novel far removed from either Baldwin's imagination/experience or that of the fictional world of Alabama and New York, were able to *do* in the body of a teenager grappling with fear and secrets, grappling with what seemed like 'taboo desires'. On the frontiers of sex and race in a place thousands of kilometres away (in a pre-digital world), reading *Another Country* gave shape and meaning to my inner turmoil. At sixteen years of age, the power of the taboo made the transgression of borders (falsely?) empowering and recognizes, as Linda Williams has written in her work on interracial sexuality in pornography, that the mere apprehension of a taboo makes it alluring *and* gives it potency (Williams 2004). Therein lie the myriad conundrums created by the borders and frontiers of race and sex.

Conclusion: the year is 2017

Unfortunately, the anthropological essentialism of nineteenth-century racial categories has not withered and died. White supremacist groups are once again in the headlines; terrorism and race are fused, to varying degrees, in the imaginations and actions of non-Muslim populations; immigration and refugee movements are accompanied by deep angst that seems to produce simultaneously right-wing politics of difference, debates about the national imaginary, and profound sorrow and melancholy on the left); the Black Lives Matter movement has (re-)energized racial politics in the United States; and in the tourist imaginary geography continues to be tethered to race in representations where Asia – the place – is synonymous with Asian peoples and culture (ditto Africa; see Bennett 2005).

In rereading Baldwin now, what is surprising is the realization that nothing in the preceding paragraph should surprise. *Another Country* is redolent with failure, self-doubt, overly acute self-awareness, secrets, trauma, haunting atmospheres and sadness. The dystopia of the modernist metropolis is writ large in Baldwin's New York, a place of alienation and loneliness, especially for African Americans, gays and bohemian whites. The bleak underbelly of the novel, I think, acted as a 'reality check' on my fantasies and romantic imaginings about the black male (homosexual) body. Already hyper-conscious of the darkness that accompanied my sexuality/sexual desire, *Another Country* was both liberating and a grim warning that nothing good could come from taboo sexual attractions.

Nevertheless, despairing and sad as the novel is, the various couplings are not judged from a moral position of absolutes. Despite the pain, the rapture, the joy and the melancholia he evokes in his characters, Baldwin never quite reinscribes the hegemonic status of whites or heterosexuals in the New York he describes. He never quite allows the dominant, powerful, oppressing patriarchal regimes of 'nation', 'white masculinity' and strictly defined 'heterosexuality' ultimately to 'win'. Rather, he creates spaces for those who are none of these. Even in despair, he seems to say, there is always the glimmer of salvation for the despairing. Above all, the despair/salvation produced in the novel by race, gender and sexuality (when the emotional effects/affects have been fictionalized with such somatic intensity) does not herald a utopia where 'race', 'gender' and 'sexuality' disappear; in fact, the very opposite.

Therefore, unsurprisingly, Baldwin's 'realism' and his extensive commentary about 'race' and racial anxieties have, in 2017, seen a renaissance. However, this renaissance – his books and essays republished; an award-winning and powerful documentary, *I Am Not Your Negro* (Peck 2016–2017) – do not and cannot address what was unleashed by a sixteen-year-old's erotic fantasy built on a teen's reading of a novel arriving, almost unbidden, from the frontiers (geographic, sexual and racial) of that youth's experience. Yes, I am intensely interested in the current reappraisal of James Baldwin's opus, but the commentary, for me, is a significant intellectual response to his voice, heard again, in the midst of the United States' current racial upheavals/perturbations. It is weighted, deservedly, towards racial politics, not homosexual desire.

What I am left with is not James Baldwin at all but a continued interest in/attraction to naked black male bodies, no longer specifically African American, but wherever African men live, whether in continental Africa or somewhere else due to the great, often painful, diaspora (slavery, colonial and post-colonial migration, refugees and other displacements) – South America, the Caribbean, Western Europe and, more recently, Australia – and the subsequent and often long settlement in places that are now properly their homes. This means the ongoing potency of borders and frontiers in my own imaginings. The racialized, gendered and erotic body remains powerfully attractive in art, cinema, media, dance, fiction and so forth. But mostly these (African) 'black male bodies' are inscribed with, and carry within, a history that is not mine. This existential difference gnaws at, critiques and, crucially, underpins sexual desire.

The contours of this vexed reality are described by Linda Williams in her 'Skin Flicks on the Racial Border' – a study of interracial lust, exploitation and pornography (Williams 2004). Building on the insights of Abdul JanMohamed (1992), Williams insists that racialized sexuality (JanMohamed's term) remains consistently about borders, border transgressions and an imaginative frontier that transcends whatever existential reality the subject occupies. Otherwise, racialized sexuality has no power or attraction, nor any sense of being taboo. While acknowledging the particular historical formations of the black male body (slavery, colonization) that still semiotically charge the hypersexualized African body, both male and female, the adoration of the black male body is also now a product of popular culture (dance, music, cinema, television, fashion, media) and sport and is a rather 'mixed bag of stereotype and fetishistic valuation' (Williams 2004: 281). Denigration, anxiety, eroticism, aesthetic beauty, fear, celebration, idealization, eroticism, envy and so on are aspects of the contemporary 'mixed bag' producing the ambiguities explored by Williams and, previously, Mercer, whose work she extols.

Another Country provided a discursive space for such ambiguity and, in a way, anticipated its current forms by decades. In the novel, bodies are gendered and racialized and are given multiple sexual identities. But in those moments when race and sexuality are affirmed as a means of giving the narrative both realism and vigour, Baldwin also challenges such categories. Eric does not just sleep with men. Rufus and Vivaldo do not just sleep with women. Partners are intentionally marked by skin colour. This produces particular social effects, invariably unsavoury and painful, but elsewhere in the novel skin colour is irrelevant in the day-to-day interactions of the characters living moment-by-moment.

The last words go, uncomfortably, to James Baldwin. He was not, of course, referring to the subject of this chapter, but his words nevertheless resonate deeply: 'I'm not a nigger. I'm a man. If you think I'm a nigger, it means you need it. And you gotta find out why' (Baldwin speaking in Peck 2016–2017). In all the years since I purchased *Another Country* and, in that time, often recalling the events/emotions this purchase set in train, I have been trying to 'find out why'.

References

Baldwin, J. (1963) *Giovanni's Room*. London: Corgi Books.

Baldwin, J. (1965) *Another Country*. London: Corgi Books.

Bennett, A. (2005) 'Whiteness'. In: D. Atkinson, P. Jackson, D. Sibley and N. Washbourne (eds) *Cultural Geography: A Critical Dictionary of Key Ideas*. London and New York: IBTauris, pp. 109–114.

Bhabha, H. (1994) *The Location of Culture*. London and New York: Routledge.

Bhabha, H. (2001) 'Unsatisfied: Notes on Vernacular Cosmopolitanism'. In: G. Castle (ed.) *Postcolonial Discourses: An Anthology*. Oxford: Blackwell, pp. 39–52.

Castle, G. (ed.) (2001) *Postcolonial Discourses: An Anthology*. Oxford: Blackwell.

De Marco, J. (1983) 'Gay Racism'. In: M.J. Smith (ed.) *Black Men White Men*. San Francisco, CA: Gay Sunshine Press, pp. 109–118.

Eguchi, S. (2015) 'Queer Intercultural Relationality: An Autoethnography of Asian–Black (Dis)connections in White Gay America', *Journal of International and Intercultural Communication*, 8(1): 27–43.

Foucault, M. (1973) *The Order of Things: An Archaeology of the Human Sciences*. New York: Vintage Books.

Gregory, D. (2004) *The Colonial Present*. Oxford: Blackwell.

Hemphill, E. (ed.) (1991) *Brother to Brother: New Writings by Black Gay Men*. Washington, DC: Redbone Press.

JanMohamed, A. (1992) 'Sexuality on/of the Racial Border – Foucault, Wright and the Articulation of Racialized Sexuality'. In: D. Stanton (ed.) *Discourses of Sexuality*. Ann Arbor: University of Michigan Press, pp. 94–116.

Lee, H. (1960) *To Kill a Mockingbird*. London: William Heinemann.

Mapplethorpe, R. (1983) *Robert Mapplethorpe 1970–1983*. Exhibition catalogue. London: Institute of Contemporary Arts.

Margulies, S. (producer) (1977) *Roots: The Saga of an American Family*. TV series. Wolper Productions and Warner Bros Television.

Mercer, K. (1991) 'Skin Head Sex Thing: Racial Difference and the Homoerotic Imaginary'. In: Bad Object-Choices (ed.) *How Do I Look? Queer Film and Video*. Seattle, WA: Bay Press, pp. 169–222.

Peck, R. (director) (2016–2017) *I Am Not Your Negro*. Documentary. Magnolia Pictures and Amazon Studios.

Shaffer, P. (1977) *Equus*. Harmondsworth: Penguin Books.

Twain, M. (1884) *The Adventures of Huckleberry Finn*. London: Chatto & Windus.

van Houtum, H. (2011) 'The Mask of the Border'. In: D. Wastl-Walter (ed.) *The Ashgate Research Companion to Border Studies*. Farnham and Burlington, VT: Ashgate, pp. 49–61.

Wastl-Walter, D. (ed.) (2011) *The Ashgate Research Companion to Border Studies.* Farnham and Burlington, VT: Ashgate.

Williams, L. (2004) 'Skin Flicks on the Racial Border: Pornography, Exploitation'. In: L. Williams (ed.) *Porn Studies.* Durham, NC, and London: Duke University Press, pp. 271–308.

2 The Spanish Imaginary

A trilogy of frontiers

Steve Watson and Emma Waterton

> My first encounter with Spain was breathing it. Like most people travelling from Britain at the end of the twentieth century I arrived by plane, with my visual sense dulled by the global ubiquity of airport architecture both inside and out. But when I emerged from the terminal to look for a taxi I breathed warm dry air, with a hint of something else, possibly fumes from the many aircraft taking off and landing, and it has stayed with me for thirty years. From the cold dampness of a Yorkshire coastal January to the crystalline sunlight warmth of Malaga, I was one of the many millions who had made the same journey from north to south in search of difference, and the difference hit me first in the lungs. Mine was thus a physiological response, a momentary gasp, remembered clearly, but never put into words until I wrote this paragraph just now.
>
> (Watson 2015)

Introducing our frontiers: geographical, conceptual and personal

There are three 'frontiers' at work in this chapter. The first regards a geographical frontier, in that the chapter takes as its focus the encounter between two highly representational cultures – the Anglo-Saxon and the Spanish–Mediterranean – in the nineteenth century, from which emerged 'the Spanish Imaginary' (after Taylor 2004; see Watson 2014). This imaginary developed over time and with the active involvement of various representational agencies, emerging as a socially constructed and imagined version of Spain that has since come to be seen as the 'real thing'. Our interest in this chapter is with the particular representational agency of British and American travel writers, who brought their own sense of modernity to look upon Spain from a perceived position of dominance – industrially, militarily and, not least, culturally. This 'Spain' has a long antecedence in the writings of British and American authors, most notably in the work of Richard Ford (1845, 1852) and George Borrow (1843) in the early to mid-nineteenth century. Since then, successive waves of travellers have reproduced the Spanish Imaginary, such as Theophile Gautier (1853), Albert Calvert (1903, 1908), Gerald Brenan (1950) and Laurie Lee (1955, 1970). Indeed, over a period of two hundred years, travel writing about Spain became the key medium through which the place, if not the nation and its identity as such, came to be known. Taking the notion of travel writing as a means of reaching into and beyond an imagined frontier, this chapter explores the writings of a particular set of authors who were active in the nineteenth century to the turn of the twentieth century: women travel writers.

In focusing upon the Spanish Imaginary, the overall thrust of the chapter necessarily takes a historical perspective. The second frontier that structures this chapter is also historical in the sense that it emerges from the experiences of women travel writers, but its focus on emotional and affective dimensions also marks it out as 'exploratory' in conceptual terms. This is because research on emotion and affect is relatively new in the field of Heritage Studies, but it is an important avenue for this chapter as the Spanish Imaginary often manifests as a sensory domain, in its atmospheres, sights, sounds and smells, and in how it feels to be there. It is also apparent in moments of absorption, illusion, seduction and fascination, and particularly when it combines with a kind of wariness and mistrust of a culture that is historically so different from that of Northern Europe and America. Many writers seemed attracted to this because it marked out a fundamental difference with their own culture, but it also repelled them in its visceral realities. In this, the chapter charts the potential for travel writers to create meaning around the places they visited, and evoke affective registers of experience and expressions of emotion to create an intense imaginary that stands in relation to their own identity and the realities of Spanish culture. These factors – the affective and emotional – provide the theoretical framework for the chapter, which intends to contribute to ever-expanding debates on emotion and affect, and find ways of applying such theories in a novel way to a country that has always been a source of interest and intrigue in the northern imagination.

While the primary source material for the chapter is the published work of women travel writers, it also draws heavily on one of the author's own experiences of Spain. So, there is a third 'frontier' to this chapter, and it is far more personal and implicit than the other two. Spanish travel writers and the Spanish Imaginary 'belong' to only one of the authors named at the start of this chapter, Steve Watson, who passed away in early 2016, before its completion. To ensure Steve's presence, I (Emma) have interspersed our shared writings with snippets from his field notes, leaving me, as co-author, caught in that living space of the chapter's becoming, crushed between Steve's presence and absence – the effects/affects of which most certainly permeate the final product.

Affecting the Spanish Imaginary: an overview

A central objective of this edited collection is to explore frontiers as spaces of transformation, exploration and adventure, and to think about the intersections between these and notions of place and identity. For this chapter, such an underpinning has prompted us to converge on the Spanish Imaginary, a place in which all of the senses are drawn together to give and reproduce those meanings by which it has come to be known. It is clear from the most cursory readings of travel writing on Spain that it is a place that has affected its visitors in different ways, their feelings registered ultimately as both deep emotions and profound thoughts. A clear example of this can be found in the work of V.S. Pritchett, whose first sight of the region of Castile changed his life. There,

> he saw Spain in its most compelling guise, where 'we learn our history upside down and see life exposed to the skin. Neither in France nor in Italy can one be so frankly frightened' (Pritchett 1954: 7). Pritchett's comments are emblematic of a type of writing about Spain that paints it in violent light and shade, bright colour and deep contrast, something not only seen but felt, a writing that draws from the experience of its authors an endless expression of difference, exoticism and a kind of Otherness that belies its western European location. And yet, like me, Pritchett was very much alive to the way he felt, not just what he thought.
>
> (Watson 2015)

Spanish travel writing, as a generalization, has always represented the Spanish Imaginary as a separate place: mythic, timeless and exotic, the source of endless explorations and discoveries that construct, sustain and reproduce a set of meanings by which Spain is made in the northern imagination. It is in the guise of this imaginary that it is anticipated by intending tourists, and sought by those who ultimately arrive there. A great deal of what constitutes this imaginary is nestled in a segment of the literature that has, as Adams (2015) points out, been long neglected by scholarly critics: the 130 or so British and American travel books published between 1700 and 1970.[1] These constitute a corpus of what has come to be known as 'travel literature' – literary, non-literary or both – and whilst they might have been used as guidebooks and even to some extent written as such, they contain enough authorial reflection and subjective commentary to make them something more than just tourist guides (Adams 2015). Inevitably, of course, there is a considerable range in the personal expressiveness of the authors' voices. In some cases, Spain seems lost to the figure of the exploring author dominating the scene; a backdrop or side-note limited by overpowering and triumphant human experiences. In other cases, the author is lost in the welter of impressions and images for which they seek expression. Travel advice is often plentiful, especially from those who saw themselves as pioneers, leading a vanguard of future travellers who must beware the predatory activities of their vagabond hosts (see, for example, Ford 1845). For the most part, however, this is a literature that is constitutive of a subjective response to place. What makes *this* genre of literature – as a response to *that* particular place – different is the set of cultural and historical factors that spin out a shared imaginary, an intersubjectivity that is jostled into shape by a unique assemblage of places, experiences, knowledge and feeling.

Foregrounding subjective registers – from the embodied and affective to the cognitive and intellectual – provides an opportunity to think about travel writing not only as a representational domain, but as one composed by experiences and engagements that are linked with the senses (our second frontier). Crucially, however, these subjective experiences are made available through the social, cultural and mechanical processes of the publishing industry, unless we can get our hands on unpublished documents, which is sometimes possible but more as a matter of happenstance than of any systematic searching. By and large, though, publication provides the material from which individual travel experience is produced, shared and made available. As such, it is the subject of mediating practices, starting with the worldview of the traveller, the ability of the traveller to write about their experiences, the influence of social and cultural discourses, the receptivity of a market place of readers, the selectivity and editorial practices of publishers, censorship and public taste, the choice-making of potential readers and the distribution and longevity of works in print. The long and circuitous route from experience to the printed page can only have a modulating effect on traveller text. We can see this in a certain congruence of features that emerges and changes over time, and which is apparent in travel writing about Spain and figured in the reciprocities that exist between the production of text and the Spanish Imaginary as it has formed over the last two centuries.

One effect of this mediating structure is that we are always, inevitably and irrevocably, separated from the experience of travellers by the very means they use to communicate it. And yet, in it we can trace something felt, understood and meaningful:

> In terms of style it matters less to me that such material is written up as literature or as reportage, than what it is intended to evoke, although I draw the line at guidebooks unless they project a clear sense of authorship. Across these various forms of expression there may be serious moments of self-reflection, philosophy, historical analysis, literary references and humour. And overlaying this matrix of form and content there are broader socio-cultural discourses reflecting colonialism, orientalism, nationalism, and the status of the individual in heavily differentiated social structures.
>
> (Watson 2015)

There is a second immediate paradox in the experience of travel, which is that it is so often the means by which to achieve a fixed position, or a set of fixed positions: the places travelled to – destinations – are defined by their qualities of place, immutable points on the surface of the earth, Cartesian promises on a map. This tension in travel writing exists between moving and remaining, and how long to remain before moving on, and what pleasure to take from the movement in itself. As Robert Louis Stevenson (1881: 4) famously reminds us, 'to travel hopefully is a better thing than to arrive'. The further back in time, of course, the more travel loomed in the experience itself, and writers make much of dealing with stagecoaches, donkeys or simply walking. It would be less of a matter to recall one's experience of the liminalities of modern railway stations and airports or driving on a motorway. The point of arriving, however, is that one arrives somewhere else, somewhere different, and it is this difference that energizes the experience of travel.

In order to understand the experience of travel, this chapter is framed by a theoretical device we originally developed elsewhere: the idea of 'encounter' (see Waterton and Watson 2014, 2015). At one level, this sense of *encounter*, understood in relation to the Spanish Imaginary, occurs between two cultures, each with its own essentialisms and deeply embedded expressiveness. But to leave it at that would be to limit the encounter to an experience of difference written in terms of imagery, national stereotypes and other banal comparisons, rather than the imagination. Certainly, those images exist and they exert a considerable influence in perception and understanding – this would be the cognitive domain, as we have described it elsewhere (Waterton and Watson 2014). But there are other ways of knowing and encountering that infuse travel writing and take the discussion way beyond the representational conventions that accrete around nations and cultures, creating a Spain that is felt and experienced through all the senses. This Spain is understood through a multimodal expressivity that ultimately construes its own imaginary based on travel as an embodied experience that is cognitively linked with knowledge received from the representational domain, including other travel writers, but which remains resolutely separate from it. The point is that the imaginary *is real enough to feel.*

In philosophical terms, this sense of encounter is associated with the politics of affect and the idea that bodily experience registered in feelings and emotions is key to understanding the body's power to act when interacting with other bodies. This understanding of affective force emerges from the work of Baruch Spinoza, and his notion that affect is the power 'to affect and be affected' (see Massumi 2015: ix). As Massumi (2015: x) goes on to argue,

> Although affect is all about intensities of feeling, the feeling process cannot be characterized as exclusively subjective or objective: the encounters through which it passes strike the body as immediately as they stir the mind. It involves subjective qualities as directly as the objects provoking them, or with which they move. It concerns desire as much as what is imperatively given; freedom as much as constraint.

While Spinoza identified forty-eight varieties of affect, they are largely derivative of three basic forms: desire or appetite; pleasure; and pain or sorrow. From the perspective of the traveller, such feelings are made readily available in the act of travelling; indeed, they are heightened by the kinds of experiences that travel may engender, either in the act of travel itself or through engagements with a particular object or destination. In the prepared spaces of travel, moments of affective engagement are both expected and provided. Thus, what is experienced is a semiotic landscape (Waterton and Watson 2014; Waterton 2018), where moments of intensity are made available and where opportunities abound for feelings to be expressed in various ways.

In his articulation, Spinoza poses a conceptualization in which affect might be considered precognitive and therefore pre-emotional – a register in the body of, for example, pleasure or pain, discomfort or satisfying a need, all of which are clearly properties of a traveller's experience. Brian Massumi (1987: xvi), whose work draws primarily from Spinoza, talks of affect as 'a prepersonal intensity corresponding to the passage from one experiential state of the body to another and implying an augmentation or diminution in that body's capacity to act'. The key feature of affect in Spinoza's and Massumi's work is this prepersonal nature, which can be understood as a register in the body that is not necessarily represented or expressed linguistically or symbolically, either to the self or to others. By contrast, emotions are by their very nature expressive and representational. This distinction has been picked up by social scientists, and particularly cultural geographers, as part of what has been referred to as the 'affective turn', and has been expressed in a canon of literature since the 1990s as 'non-representational theory' (see Thrift 2008).

The introduction of non-representational theory – or more-than-representational theories, as the approach has more recently been termed, after Lorimer (2005) – instigated a paradigmatic transformation in thinking about the way people construct meaning around their subjective lived experiences. In particular, adopting such an approach prompted a focus on the performativity of experience and *meaning in doing* rather than the mapping of experience across precast representational categories that are emotionally expressive or socially constructed. As the locus of such registers is the body itself – rather than the conscious, prescient mind making sense of the world through which it is moving – affect would seem central to the experience of travellers. The essence of travel, after all, is the movement from one set of environmental conditions to another. Indeed, its purpose is to experience the differences that are engaged in such movements, and those differences are registered sensually before they are represented in other ways. Travellers, therefore, experience the world primarily as bodies, and such experiences cannot be disembodied (Lean *et al.* 2014). The body in motion and the emphasis on the sensory immediately contextualize the visual in a broader sensory array. This is ironic, given the privileging of visuality in tourism, where sightseeing has been seen as *the* central activity in the culture of travel since at least the eighteenth century (see Waterton and Watson 2010). Tourists and travellers seek views, which are often conventionalized in guidebooks, postcards and more contemporary media. In some locations there are 'authorized viewpoints' that seem to beckon us to them to complete the ritual of sightseeing by taking a photograph, thus honouring once again a particular image (Waterton and Watson 2014). But more-than-representational theories provide a conceptual basis for de-privileging the visual and giving due emphasis to the other senses in embodied and performative experiences and engagements with objects and places. As Waitt and Macquarie (2014: 53) have argued:

> The shift from representational to embodied approaches acknowledges that people do so much more when they travel than only read guide books, advertisements, go on sightseeing tours and take photographs. Places are never sensed and made sense of by sight alone.

There is no doubt that such more-than-representational registers are a property of all travel and indeed of movement in general, including the movement from one room to another in one's own home, or in the immediate world of a bounded or constricted environment engaged by the *bodied* self (Bell 2014). In what follows, we will make a case for the existence of a specific affective register that is evident in travel writing about Spain. The specificity of this register is, perhaps revealingly, as much an artefact of prior representations of Spain as it is of direct experiences of Spain itself, and this is crucially important in our understanding of the Spanish Imaginary.

Histories of Spanish travel writing

> The act of travel writing charts a range of expressive capacities from the sensual and essentially pre-expressive through layers of increasing expressiveness, knowing and articulation. This much would be true of any travel writing, but what interests me about Spain is the extent to which this process of meaning making is essentially and centrally reflexive, constantly feeding upon itself, reproducing and amplifying its core sentiment with each iteration to create a discursive imaginary wherein another Spain is constructed and sustained, separate from its quotidian existence, but inextricably linked with it, influential, pulsing.
>
> (Watson 2015)

Histories of travel writing on Spain can be found in two excellent anthologies by David Mitchell (2004) and Alice Leccese Powers (2007). The former, *Travellers in Spain: An Illustrated Anthology*, is organized around authors commenting on Spain across four centuries, while the latter, *Spain in Mind: An Anthology*, offers a selection of fiction, non-fiction and poetry from English and American writers spanning three centuries. While it is not our intention to cover the same ground in the same way as these excellent contributions, the historical narrative they afford is an important part of our account. It begins with the assertion that Spain was 'discovered' by writers bringing their own sense of modernity to its shores from Britain and the United States. While the latter was at least sentimentally a liberated colony, both nations looked at

Spain from a position of dominance. The United States had quickly disposed of the remains of Spain's empire in Cuba in 1898, simultaneously sparking an expansion of America's overseas possessions. This is a moment captured in the celebratory tone of J.B. Crabtree's (1898) *The Passing of Spain and the Ascendency of America*. For Spain, it was a shocking debacle and an ignominious footnote in the epitaph to its empire. This was the degeneration that marked a great deal of the difference between its own experience and that of the emergent modern industrial nations that watched it through the eyes of travellers.

There are many features of Spanish history and culture that stand out as emblematic and essential, but the overriding one is this general sense of difference and 'otherness'. The sources of this have been the stuff of travel writers since the early nineteenth century and are exemplified in a number of recurrent themes: the Moorish heritage and its traces found in buildings and people; Gypsies and flamenco; a starkness in the landscape in contrast with the brightness and vivacity of its cultural productions; the bullfight (almost without exception); the built heritage; the obdurate quirkiness of the people (frequently linked to their Moorish heritage or their ascribed atavism); a feeling of decline; the Civil and other wars; and, of course, the climate of the south. The experience of Spain for those whose cultural origins might be described generally as 'northern', whether British or North American, begins therefore, with the experience of difference. But there is more to its Otherness and difference. There is in Spain a deeper sensuality – something felt, something intense – that seems lodged in places and yet mobile in the fluid engagements of the traveller and expressive in their imagination and writings:

> It has become my objective to understand the way that difference is felt, constructed, expressed and then reiterated time and time again in broader thinking about culture, travel, encounter and engagement. This involves coaxing a little theory from sources that have previously explored the representational and non-representational, along with the nature of encounter and engagement in cultural and spatial domains. It may even involve testing such theory, stretching it a little, pushing it around. And there are opportunities to do this in the experience of difference. I thought of some words of Kathleen Stewart's at this point, thinking about Spain and what it is that our generalized northerner finds in it: 'we wish for the simple life that winks at us from someone else's beautiful flowerbeds' (Stewart 2007: 10). And what flowerbeds Spain has provided!
> (Watson 2015)

The Spain of travel writers is thus a curious mixture of experience and imagination, and never a straightforward exploration of the unknown or unfamiliar. Indeed, this is a realm where the familiar and the exotic collide, sometimes alarmingly, and where, paradoxically, the unfamiliar is well known, celebrated and actively sought. The source of this paradox is the contrast between its European cultural geography and its marginality in that culture. Spain is somehow on the edge of the known world and, at the same time, connected with both the classical origins of Europe and its greatest threats to the east and the south. The notion of 'Otherness' in this context is, of course, central to its appeal for travellers and an abbreviated orientalism is found in much of the writing. We should remember, however, that while orientalism is a property of those who write about Spain rather than Spain itself, there has been a curious transfer of this quality so that it came to be sought as a defining characteristic of place and people, regardless of how they saw themselves. It could be argued that such experiences, especially where we might term them 'exotic', are readily available elsewhere in the world. Asia, for example, has for centuries provided alluring combinations of climate, landscape and cultural difference to fire the Western imagination. But Spain, whilst exotic – at least by attribution – is 'here', not 'there'. For our travel writers hailing from Britain it was only a few days of travel away. Even as far back as the eighteenth century, the heartlands of oriental Andalucia were only a stone's throw from English Gibraltar, or Cádiz, as a gateway port receiving sailing ships and steamers from Liverpool, Glasgow and across the Atlantic.

The Spanish Imaginary: women as travel writers

Marginality is a condition of both women as travellers and of Spain as a European country. Despite its proximity to Britain and Northern Europe, Spain has always had the feeling of marginality, even extremity, as a peninsula jutting out not into the Mediterranean habitus of Greece and Rome, but beyond the Pillars of Hercules and into the Atlantic, yet within sight of Africa. This proximity with Africa has, of course, played a determining role in Spain's history. The Moors, under the Umayyad Dynasty, invaded the peninsula in 711CE; one of their generals, Tariq ibn Ziyad, gave his name to Gibraltar (Jebel Tariq) in passing. The invasion was as swift as it was unremarkable, as Bernard Reilly (1993: 51–52) has put it, given the long-established connections between the two continents that their physical closeness engendered (the Straits of Gibraltar are less than twenty kilometres across). The significance of the Moorish heritage in Spain is obvious, but in a sense it has an ancillary quality. Spain would have felt exotic to Northern Europeans even without its Arabic adornments. Spain would always, regardless of its orientalism, have been a place of affective engagements. The Moors simply created and rationalized a route to those feelings and provided one way of expressing and describing them. Travel writers provided another, but they had the Moors to act as a cipher, an orientalized abbreviation, as Said (1978) would have it, of everything that was different. Severn Teakle Wallis, a Baltimore lawyer and Hispanist who travelled in Andalucia in 1847, made the point well:

> The reader is not to suppose, for an instant, that Seville is a mere depository of Moorish relics and monuments of old; for, on the contrary,

> its men, women, and children are as full of vitality and spirit and the present time, as if they had never heard of hoar antiquity.
>
> (Wallis 1850: 176)

So, there is something about Spain that registers in northern travellers a capacity to be affected, but it is not always due to the conveniently oriental aspect that the Moors provide. Indeed, the 'Otherness' of Spain also has a gendered quality. It is possible in much of the travel writing on Spain to detect an expression of northern power over southern weakness; a masculine, muscular, ordered, industrial 'north' that is unsettled – seduced – by a Spain that is fey, unpredictable and occasionally violent:

> Yet the old Arabian enchantment casts a glamour which the Anglo-Saxon vision dimly recognizes as such and faintly strives against. To the clear survey all is not charm. Grace, mirth, and music, on the one hand, are offset by ignorance, suffering, and vice on the other. Many evil things were told us, and some ugly things we saw, but to look on Andalusia is to love her, even while realizing that to live with her would put that love to a very stringent test.
>
> (Bates 1900: 52)

In adopting a gendered perspective in this chapter, our interest is with the role women travel writers have played in relation to the creation and dissemination of the Spanish Imaginary. As we argued earlier, the thing that enabled Spain to be so effectively 'Othered' was its Moorish heritage, which is something that travel writers referred to constantly. The Gypsies were useful, too, in that they presented something savage and elemental, and thus ideal from a colonial perspective – a reason for disparagement, then a source of superiority by contrast and finally an implication of the need for control. The question that necessarily arises from this mix, however, is whether, in the extent to which they are women contributing to and reflecting the Spanish Imaginary, women acted as colonists or did they reflect the experience of colonialism in their own marginalization? It gets even more complex when we ask about the relationships between women as either colonizers or marginalized and the colonized/orientalized natives of Spain. In what follows, we aim to introduce readers to an eclectic range of British and American women travel writers before turning to address these questions.

> Women have been represented throughout the period and, indeed, the earliest was actually one of the first people that I can reasonably describe as a travel writer. This was Madame D'Aulnoy ,whose book, published in Paris in 1692, has the delightful if less than snappy title of *The Ingenious and Diverting Letters of the Lady: Travels into Spain; Describing the Devotions, Nunneries, Humours, Customs, Laws, Militia, Trade, Diet, and Recreations of that People.* Madam D'Aulnoy also wrote fairy stories,

> including ‘Goldilocks’, ‘The Wonderful Sheep’ and ‘The Frog and the Lion Fairy’. I am sad to have to relate, however, that her travel book was also a fairy story, there being no evidence that she ever set foot in Spain. Most of her content was obtained by a method that we know today as plagiarism but which in those days was known as exceptional talent!
>
> (Watson 2015)

Notwithstanding Madame D’Aulnoy’s brief and questionable foray into travel writing, one of the earliest women to write about Spain was Lady Elizabeth Vassall Fox, who travelled through Europe in the 1790s before journeying to Spain in 1802–1805 and 1808–1809 (and thus coinciding with the commencement of the Peninsular War). By most accounts, she was a formidable and flamboyant woman, famous for her rudeness and wit, and marginalized only in the sense that she was a divorcee (Colley 1992). Her writings betray a strongly colonial sentiment in her views on Spanish men and women:

> The women are ungraceful out of their mantilla; allow their voice to get into high tones, but seem to enjoy conversation, which they enliven frequently with sallies of humour and even wit. Unlike the Italians in many respects, they resemble them in that making love is the sole occupation of their lives.
>
> (Vassall Fox 1910: 19)

While she may have been talking about any other colonized people, a caricature of the Spanish nonetheless emerges, one that characterizes them as obsessed with their appearance and in which their high manners hide a lascivious nature. These are people, according to Vassall Fox, who are largely uneducated. Seville is described as abominably hot and offering no repose, night or day, while the inevitable bullfight is viewed as inhuman and barbarous, a practice from which she withdraws in genuine disgust, wishing a just retribution on the brutal picadors and matadors. All the elements of the imaginary are here and seen through the eyes of superior sentiments. She was oppressed by the heat, fascinated by the architecture, observant about the conditions of the people, the beggars and the filth in Córdoba – even in the cathedral – yet sanguine about much of what she encountered, apart from the bullfight.

Vassall Fox was followed by many other formidable British characters:

> Mrs Ramsay published *A Summer in Spain* in 1874. I can find out absolutely nothing about her except that she translated Dante’s *Divine Comedy*, she was British and she lived in Rome. Here we have an indomitable matriarch, sorting things out, pressing on, despite the fact that she crossed the border with her family right in the middle of the Carlist uprising. She did, however, meet her match in a hotel in Burgos, when she awoke to find she had shared her room with most of Spain’s

> extravagantly varied insect population. This woman could have traversed Australia with three children and nothing more than a packet of biscuits; she could have quelled the Indian Mutiny by having a quiet word; but it was in Granada that she found the imaginary she had so obviously been seeking, inspired by Washington Irving, in the cool chambers of the derelict Alhambra.
>
> (Watson 2015)

In the last quarter of the nineteenth century, American women appear for the first time. Two, in particular – Kate Field in the 1870s and Susan Hale in the 1880s – provide the first accounts of family holidays as opposed to travel adventures. Shortly thereafter, a most interesting American writer appears – Katherine Bates – whose *Highways and Byways of Spain* (Bates 1900) was published in New York at the turn of the twentieth century. Bates was a deeply committed Congregationalist, a songwriter, a social reformer and very interested in women's educational issues; and she clearly found Spain a challenge. She wrote a great deal about bullfighting in her short book, having sampled a contest in San Sebastián almost as soon as she arrived, but she saves her best prose for the Easter Day bullfight in Seville. Here, she plays well the role of the judgemental colonist – bullfights, of course, provide an opportunity for claiming the moral high ground, although she does provide a perfect account of the embodied, emotional and cognitive aspects of the experience.

The remainder of the twentieth century, by contrast, saw less colonial sentiment and more reflective accounts of the engagements of women with Spain as travellers rather than representatives of superior powers. In 1930, Alice Riley, a well-known American writer of children's books, set off from Gibraltar in an imported Packard Series Six touring car driven by a Spanish chauffeur. Her book, *Skimming Spain in Five Weeks by Motor* (Riley 1931), was published in Los Angeles the following year. The volume was intended to be a pleasurable excursion in an Arcadian paradise, and concluded with the following advice: it was possible for young women to travel by motor car around Spain as long as one of them could speak Spanish and they kept a well-stocked luncheon basket and stayed in luxury hotels. Riley was swiftly followed by Clara Laughlin, born in New York in 1873, who grew up in Chicago and became a journalist and radio broadcaster in the 1920s, specializing in travel advice for women. One of her 'So you're going to …' series of books, *So You're Going to Spain* (Laughlin 1931) comprised the century's first truly useful Spanish travel guide. It offered a very personal and densely written account of a journey, or journeys, and nowadays we might wonder why it was written as a guidebook rather than a volume of travel literature.

Laughlin's work was the first signal of a shift away from colonial sentiment to alternative responses to Spain. A clear pattern in these later works is that of the woman writer whose relative freedom in the north endowed her with a special kind of position in 'backward', medievalist Spain. There, the freewheeling woman emerged as an icon of Western modernity, once again defining that modernity by contrast with the 'backwardness' of the Moor indigene. There was also nostalgia for an imaginary Spain under threat, particularly as the 1930s wore on and the prospect of civil war increased. A minor novelist, Gertrude Bone, was travelling in Spain at this time with her husband, Sir Muirhead Bone, an artist of the First World War, who provided the illustrations for her book, *Days in Old Spain* (Bone 1938). This elegiac work, written at the height of the Civil War, chronicled a country that Bone had visited many times before but which she now saw destroying itself.

Perhaps the best pre-Second World War book, however, is Kate O'Brien's *Farewell Spain* (O'Brien 1937), published in London and illustrated with simple, beautiful drawings by Mary O'Neill. O'Brien was an Irish novelist and journalist who had worked at the Manchester *Guardian* in the 1920s, and she produced one of the most visceral, emotional and analytical accounts of the death of old Spain at the hands of Francisco Franco. The book was written in part to raise funds for the leftist cause but also because she thought her story of Spain's suffering would act as a warning, and thereby prevent further conflict in Europe. Of course, that proved to be a vain hope.

A hiatus in women travel writing commenced with the start of the war and continued for ten years; it ended with Rose Macaulay's monumental *Fabled Shore: From the Pyrenees to Portugal* (Macaulay 1949). A Victorian lady in her mid-sixties, Macaulay drove unaccompanied through Spain with the eyes of a classical scholar detecting the residues of Greece and Rome through more recent and substantial layers, not least the debris of the Civil War:

> To my mind, Macaulay is the most successful of the authors I have consulted, in that she goes beyond journalism or the discipline of the academic into the seamless blend of the cognitive and the affective that marks all great travel writing. She was immersed in what she sensed as well as what she merely noticed; her experiences in, and of, Spain were embodied as well as intellectual, sensual and engaged, as if what she relates is newly emergent in every encounter. She swims in the sea whenever the opportunity occurs and presents such experiences alongside her academic reflections on a church or a ruined castle. Although Macaulay was from a previous generation her writing seems oddly timeless.
>
> (Watson 2015)

Macaulay was followed by Madelaine Duke, a writer of spy and detective fiction who has now more or less disappeared off the radar, and Lilbush Wingfield, writing as Honor Tracy, a British novelist and journalist who began travelling after the Second World War as a special travel correspondent for the *Observer*. Unsurprisingly, Duke (1957), in *Beyond the Pillars of Hercules*, writes like a novelist, with a surfeit of well-remembered dialogue and clearly inscribed characters. Tracy, on the other hand, writes sparely, like an extremely good journalist, with perception and great wit. *Silk Hats and No*

Breakfast: Notes on a Spanish Journey (Tracy 1957) was followed by two more books in a similar vein – *Spanish Leaves* (Tracy 1964) and *Winter in Castille* (Tracy 1973).

Given the diversity just described, it seems pertinent to ask what these women had in common in the way that they responded to the imaginary and the extent to which they reproduced and sustained it. With regard to the imaginary, Madelaine Duke made it very clear that she was prepared for the differences she expected to encounter. Travelling by rail to Ronda and Córdoba in 1957, she recounted an encounter with a young man who watched admiringly as her travelling companion used a makeshift device to remove the cork from a wine bottle: 'He said he was descended from the Moors, and he looked it. He had the sinewy pliant body of a desert Arab, a dark skinned face with a curved hooked nose and quiet, heavy lidded eyes' (Duke 1957: 124. Honor Tracy wrote in similar terms about her visit with Gerald Brenan, the doyen of the Spanish Imaginary, who in later life set up home in a village called Churriana near what is now Malaga airport. Back then, it was a quiet and idyllic spot between the city and a quaint fishing village called Torremolinos, where Tracy went swimming with Brenan and his wife Gamel Woolsey. As they attempted 'to undress decently in the shelter of a fishing boat', a youth approached and stared at them 'such as, outside of the Orient, is to be experienced only in Spain' (Tracy 1957: 27). She continues, '[T]he face is vacant, the mouth hangs slightly open and the whole power of the creature is concentrated in the two eyes that drink in the spectacle before them, unconscious of anything else and dead to shame' (Tracy 1957: 27).

This aspect of the imaginary is common in much of the travel literature: the Spanish exhibiting their 'Otherness', backwardness, orientalism and Moorishness. Rose Macaulay complains of it in Catalonia and, later, in Tarifa: '[T]here is only one drawback for foreigners on these lovely shores, as elsewhere in Spain, the inhabitants stare and point' (Macaulay 1949: 29). She also remarks on what she sees as the risk of being mobbed by 'stalking savages' and 'Arab-eyed groups of natives' who target female foreigners in an odd mixture of harassment, begging and mockery. Their interest, she writes, is 'in women, and more particularly, in women alone' (Macaulay 1949: 30). We see here a particular form of the imaginary, forged in the experience of women. They reflect and sustain the imaginary in their perception of themselves as modern women in a primitive land, at risk and yet basking in their superiority over the savage, non-European mind that is still to be found in Spain. For those who travelled south, there was often a very clear sense of descent, or perhaps immersion, measured by the growing Moorish influences south of Valencia, where Macaulay's expectations were more than met: 'Everywhere are Moorish towns and villages, place names, mosques, minarets, irrigation channels and watermills. The language has Arab words, much of the music and song is African' (Macaulay 1949: 80).

The Moors were 'the Orient' and were thus sought out everywhere, as if in these vestiges, these tokens of long-gone exotic Otherness, there are talismanic representations of difference, and of Spain itself. This was a constant preoccupation, especially where Moorish connections could be made in accounts of historic art and architecture, but also rather questionably in the look of the people, especially in Andalucia, which in the terms of the imaginary had almost come to stand for Spain as a whole. For Jan Morris, author of *Spain*, Spanishness itself is 'a pungent substance that floats around you the moment you cross the Spanish frontier' (Morris 1964: 31). This affective response was always conditioned by the experience of polarities that almost define the imaginary: familiarity and strangeness, proximity and Otherness; modernity and pre-modernity; beauty and violence; exoticism and dirt; the wealth of culture and the poverty of the people – the 'killing poverty', as Honor Tracy (1957: 16) called it. To this extent, these women were reflecting the imaginary and in their writing they were sustaining and reproducing it. In a more affective sense, they were also *feeling* it in the way they sought to understand that difference by placing it in historical, cultural and what nowadays would be unacceptably racial contexts:

> During the twentieth century these women were conscious of a difference in the way they were treated in Spain as compared to their northern homelands, and they used this to underscore, by contrast, their own modernity, a reflection of that other orientalist tendency to define by contrast. This is a curious convergence and perhaps one I am at risk of overworking but I find it quite compelling. Here were women travelling alone, driving cars, being journalists, being stared at, enjoying risk and adventure, defining their own modernity by contrast with the Spanish women they met and by contrast with Spain itself; and all of this in parallel with the defining of modernity in northern Europe by its contrast with the backwardness of Spain, a true reflection of Said's concept of Orientalism.
>
> (Watson 2015)

But there is another convergence at work here, one that is subtler but equally compelling. How easy was it for women to travel alone in Britain and France in the 1930s, 1940s and 1950s without notice or comment? They may have had more freedom, but this was obviously only a relative position. Their own feeling of modernity may therefore have been self-imputed, and *enabled* by the perceived backwardness of Spain. These women were immensely successful and their very success marked them out. They were unusual to the extent that they occupied the margins of their own society, not its centre. Their careers, activities, attitudes and behaviour singled them out, and some of them, such as Honor Tracy and Rose Macaulay, achieved great celebrity as journalists and authors. This other convergence, then, is the common experience for both women and 'Spain' to be on the margins. And in that marginality they found a subtle expression of identity and the power of self-definition: a woman writer, a lone woman traveller, in Spain – that country on the margins of Europe, dark, violent, oriental and alluring. So, while the power of definition belongs to the north, and to men, and while the state of otherness and marginality is not sought but imposed, it can also be used by the marginalized as a source of distinction and identity. These women, like Spain itself, were not merely marginal. They were different.

Concluding thoughts

One of the essential characteristics of affect, as we approach it in this chapter, is its dislocation from the conscious subject, its unconscious transmissibility between bodies and intersubjective sensing. Crucially, then, we not only reach down and within the subject to find the deep roots of affect but also outwards and into the physical and social environment, the intersubjective sphere in which a body is in constant interaction with its surroundings. In this chapter, we have pointed to a level of intersubjectivity that is evident in travel writing and transcends individual expressiveness, patterning itself across the genre or sub-genre that is Spanish travel writing. It is from this patterning that the imaginary described here is ultimately drawn. In other words, while individual experience is always constrained by the words that are available to express it, there is a collective expressivity that in its totality points to the existence of a shared cultural 'space'. As Steve so eloquently put it in his field notes:

> If we step back from the individual subjective account and look at the intersubjective sphere of expression, is it possible that we might find a place of imaginative encounter that is infused with representation, performance, language and experience? We do, after all, live in a cultural world, a world of representations that exists in reciprocity with experience, a world where the imagination itself binds the two together. But there again, this is to privilege once more the linguistic in making it the sole repository of meaning constructed from experience. There is a remove, a separation from the pre-cognitive, pre-expressive domain of embodied experience, and yet ultimately this experiential and multimodal expressivity is funnelled through language into representation.
>
> (Watson 2015)

It is, indeed, as if language – in this case the language of travel writers – is but a thin yet impenetrable crust that occludes the very experiences that feed it and that, paradoxically, it seeks to convey. Can there be any way out of this representational web?

What became of our Spanish Imaginary? In time, the marginality of Spain disappeared. The cheap package holidays of the 1950s and 1960s saw to that. And while the imaginary lived on in wry comments about '*mañana*' and waiters called Manuel, it became an object of the tourist rather than the traveller. 'Spain is Different' became the advertisers' tagline in Franco's tourism development campaign, yet the more its difference was institutionalized, the less different it became. Nowadays, the imaginary is to be found in 'Real Spain' and the urge to draw more upmarket tourists into the cultural destinations that lie inland and away from the beaches. In quiet towns like Medina-Sidonia and in the gardens of the Alcázar in Córdoba you can still find its echoes, but the women travellers who framed, sustained and reflected it in their own experiences are, of course, long gone.

Note

1 For an extensive list of these works, consult the Bibliography of Spanish Travel Writing prepared by Steve Watson and available at: www.yorksj.ac.uk/schools/york-business-school/research/futures/havrc/havrc-publications–research/.

References

Adams, P.G. (2015) *Travel Literature and the Evolution of the Novel*. Lexington: University Press of Kentucky.

Bates, K.L. (1900) *Spanish Highways and Byways*. London: Macmillan and Co.

Bell, H. (2014) 'The Jealous Imagination: Travels in My Mind and out of My Body'. In: G. Lean, R. Staiff and E. Waterton (eds) *Travel and Imagination*. Farnham: Ashgate, pp. 39–52.

Bone, G. (1938) *Days in Old Spain*. London: Macmillan.

Borrow, G. (1843) *The Bible in Spain; or, the Journeys, Adventures and Imprisonments of an Englishman, in an Attempt to Circulate the Scriptures in the Peninsula*. London: John Murray.

Brenan, G. (1950) *The Face of Spain*. London: Turnstile Press.

Brennan, T. (2004) *The Transmission of Affect*. New York: Cornell University Press.

Calvert, A.F. (1903) *Impressions of Spain*. London: Philip, Son and Nephew.

Calvert, A.F. (1908) *Granada, Present and Bygone*. London: J.M. Dent & Co.

Colley, L. (1992) 'Women and Political Power', *Wilson Quarterly*, 16(2): 50–58.

Crabtree, J.B. (1898) *The Passing of Spain and the Ascendency of America*. Springfield, MA: The King Richardson Publishing Co.

Duke, M. (1957) *Beyond the Pillars of Hercules: A Spanish Journey*. London: Evans Brothers.

Field, K. (1875) *Ten Days in Spain*. Boston, MA: J.R. Osgood and Company.

Ford, R. (1845) *A Hand-book for Travellers in Spain, and Readers at Home: Describing the Country and Cities, the Natives and Their Manners; the Antiquities, Religion, Legends, Fine Arts, Literature, Sports, and Gastronomy: With Notices on Spanish History*. London: John Murray.

Ford, R. (1852) *The Spaniards and Their Country*. New York: G.P. Putnam.

Gautier, T. (1853) *Wanderings in Spain*. London: Ingram, Cooke and Co.

Hale, S. (1883) *A Family Flight through Spain*. Boston, MA: D. Lothrop and Co.

Laughlin, C.E. (1931) *So You're Going to Spain*. London: Methuen and Co.

Lean, G., Staiff, R. and Waterton, E. (2014) 'Reimagining Travel and Imagination'. In: G. Lean, R. Staiff and E. Waterton (eds) *Travel and Imagination*. Farnham: Ashgate, pp. 9–22.

Lee, L. (1955) *A Rose for Winter: Travels in Andalusia*. London: Hogarth Press.

Lee, L. (1970) *As I Walked out One Midsummer Morning*. London: World Books.

Lorimer, H. (2005) 'Cultural Geography: The Busyness of Being "More than Representational"', *Progress in Human Geography*, 29: 83–94.

Macaulay, R. (1949) *Fabled Shores: From the Pyrenees to Portugal*. London: Hamish Hamilton.

Massumi, B. (1987) 'Notes on the Translation and Acknowledgements'. In: G. Deleuze and F. Guattari, *A Thousand Plateaus*. Minneapolis: Minneapolis University Press, pp. xvi–xix.

Massumi, B. (2015) *The Politics of Affect*. Malden, MA: Polity Press.

Mitchell, D. (2004) *Travellers in Spain*. Malaga: Santana Books.

Morris, J. (1964) *Spain*. New York: Prentice Hall Press.

O'Brien, K. (1937) *Farewell Spain*. London: William Heinemann.

Powers, A.L. (2007) *Spain in Mind*. New York: Vintage Books.

Pritchett, V.S. (1954) *The Spanish Temper*. London: Chatto & Windus.

Reilly, B.F. (1993) *The Medieval Spains*. Cambridge: Cambridge University Press.

Riley, A. (1931) *Skimming Spain in Five Weeks by Motor*. Los Angeles, CA: Saturday Night Publishing Company.

Said, E. (1978) *Orientalism: Western Conceptions of the Orient*. London: Penguin Books.

Spinoza, B. de (2001) [1677] *Ethics*. Trans. by W.H. White and A.H. Stirling. London: Wordsworth Editions.

Stevenson, R.L. (1881) *Virginibus Puerisque: And Other Papers*. London: Chatto & Windus.

Stewart, K. (2007) *Ordinary Affects*. Durham, NC: Duke University Press.

Taylor, C. (2004) *Modern Social Imaginaries*. Durham, NC: Duke University Press.

Thrift, N. (2008) *Non-representational Theory*. London: Routledge.

Tracy, H. (1957) *Silk Hats and No Breakfast: Notes on a Spanish Journey*. London: Random House.

Tracy, H. (1964) *Spanish Leaves*. London: Random House.

Tracy, H. (1973) *Winter in Castille*. London: Eyre Methuen.

Vassall Fox, E. (1910) *The Journal of Elizabeth Lady Holland (1791–1811)*. London: Longmans.

Waitt, G.R. and Macquarie, P. (2014) 'Travel-as-Homemaking'. In: G. Lean, R. Staiff and E. Waterton (eds) *Travel and Imagination*. Farnham: Ashgate, pp. 53–71.

Wallis, S.T. (1850) *Glimpses of Spain; or, Notes of an Unfinished Tour*. New York: Harper and Brothers.

Waterton, E. (2018) 'Curating Affect: Exploring the Historical Geography–Heritage Studies Nexus at Sovereign Hill', *Australian Geographer*, 49(1): 219–235.

Waterton, E. and Watson, S. (2010) 'Introduction: A Visual Heritage'. In: E. Waterton and S. Watson (eds) *Culture, Heritage and Representation: Perspectives on Visuality and the Past*. Farnham: Ashgate, pp. 1–16.

Waterton, E. and Watson, S. (2014) *The Semiotics of Heritage Tourism*. Bristol: Channel View.

Waterton, E. and Watson, S. (2015) 'The Ontological Politics of Heritage; or How Research Can Spoil a Good Story'. In: E. Waterton and S. Watson (eds) *The Palgrave Handbook of Contemporary Heritage Research*. Basingstoke: Palgrave Macmillan, pp. 21–36.

Watson, S. (2014) 'Making Spain: The Spanish Imaginary in Travel Writing since the Second World War'. In: G. Lean, R. Staiff and E. Waterton (eds) *Travel and Imagination*. Farnham: Ashgate, pp. 119–135.

Watson, S. (2015) Unpublished field notes.

3 Under a volcano

A journey through ruin and regeneration in Bali

Denis Byrne

Introduction

In March 1963, it took three days for word of the eruption of Gunung Agung to filter through to the *Straits Times* in Singapore. Dominating the eastern end of Bali, Agung was the island's largest volcano. An image of its almost perfectly symmetrical cone hovered in my mind as I reeled through pages of that newspaper on a microfilm reader in the Fisher Library at Sydney University in early 1994. The first reference to the eruption was in the form of a Reuters report for Wednesday 20 March, filed from Jakarta and based on information gleaned from 'travellers reaching here' ('fleeing here' might have been more accurate). One of those travellers happened to be the chief editor of *National Geographic* (Booth 1963). He reported that he had been driving through the Balinese countryside on the morning of the eruption, enjoying the tranquil scenery, when he became aware of a tapping sound on the roof of his car. He got out to find volcanic ash and cinders falling out of a darkening sky.

Ash began falling on the island of Madura and on 18 March clouds of it blacked out the sky over Surabaya and then Jakarta. When correspondents reached Bali some of them chartered light planes and flew with their photographers over the lava flows and the destroyed villages and reported a 'mushroom cloud' standing over the mountain. This was 1963, when nuclear testing had not yet gone entirely underground and the mushroom cloud was a symbol of both post-war modernity and the appalling events that had brought the war in the Pacific to an end only eighteen years earlier. There followed first-hand accounts from those correspondents who had penetrated the disaster zone on the ground.

After nearly half a century of mythologizing Bali as an Oriental island paradise, few people in the West had not read about the place or seen pictures of it (Vickers 1989). The press now played on the pathos of the situation: the kindly, peaceful inhabitants of an island paradise suffering in the wake of a natural disaster of a kind that was itself only too exotic. The eruption had 'made a wilderness of a paradise', the *National Geographic* observed, and in the process had 'killed more than 1,500 of some of the world's handsomest and most intelligent people' (Booth 1963: 436). In the midst of the Cold War, would the magazine's Western readers have been less distressed, one wonders, if they had known how many Balinese were communists?

Deposition

Early one morning in July 1994, having returned to Bali, where I had been living for the previous two years, I locked the bamboo garden gate of my bungalow in Seminyak and drove towards Candi Dasa (the 'C' pronounced 'Ch'), the small beach resort on the south-east coast of the island where the lower, gentle slopes of Gunung Agung reach the sea. Situated on the far side of a long bay, Candi Dasa was just a small and remote fishing village in the early 1960s, and in the 1990s it was still a sleepy place even in the high season – those months of dry and relatively cool weather following the end of the monsoon. Its bungalow hotels were squeezed into the narrow strip of flat land at the foot of the parched hills and they looked out to a sea which saw a light traffic of outrigger sailing canoes and occasionally, further out, an inter-island trading *prahu*. Young travellers gathered at night at one or other of the pavilion cafés to watch videos that were advertised on handbills plastered over the trunks of coconut palms along the sandy lanes and paths. The music and voices were swallowed by the darkness and diminished by the sound of the sea. And on the landward side there was the dark backdrop of the hills and of Gunung Agung itself, whose presence on a moonless night you would sense more than see.

I had come to the island's eastern end to look for traces of the 1963 eruption, but not in a systematic, still less scientific, way (see Figure 3.1). I wanted to think about the signature of loss in the landscape. Volcanic eruptions erase, more or less, the familiar, intimately inscribed landscape of everyday life, like a duster drawn across a blackboard full of closely written text. And yet perhaps this is not a good analogy. The board is never wiped clean and left empty; the act of erasure itself leaves myriad traces or signs which themselves may be eloquent reminders to people – the survivors, the bereaved – of their loss. Archaeology, you could say, abhors a vacuum: the blackboard is never left empty.

The day after arriving in Candi Dasa I drove further east, taking the winding road up and over a steep, bush-covered ridge, one of the spines of high ground radiating out from Gunung Agung. I passed a brick spirit shrine near where the road crested the ridge and a stall selling watermelons halfway down the other side, then drove across the flat floor of a valley through an irregular grid of irrigated rice fields. I passed through a village where the house compounds had drystone walls, and it was just on the other side of this that I noticed, to the left of the road, the remains of a bridge, ten metres upstream from the small bridge I was crossing. I pulled over and walked back to it. There were the masonry stumps where the bridge had connected with the bank on either side and, leading away from them, strips of old tarmac

Figure 3.1 The beach at Cadi Dasa, 2017
Photograph by Denis Byrne

marked the route of the road as it had been in 1963. Further on, in the Regency of Karangasem, I passed a couple more of these phantom bridges that had fallen victim to the eruption thirty years earlier.

A few kilometres past the remains of the bridge the road veered inland along a broad valley until it came to the village of Subagan. The newspapers in March 1963 had carried photographs of a street in the village where shops and houses were half submerged below a deposit of ash and volcanic rock which had settled around them, spilling in through doors and windows. Torrential rain had fallen on the higher slopes of Gunung Agung, loosening the accumulation of hot ash and rock that had settled there and bringing it down the mountainside in a steaming torrent more than half a kilometre wide. The torrent had gradually slowed in momentum until, when it reached Subagan, instead of carrying the village away, it had, as it were, filled it up. But it was no less lethal for that. Two hundred people praying in a temple at the time it struck were killed. The Malay word *lahar* describes these mobile, rain-saturated deposits of hot ash and rock, which, in that part of the world, are considered to be far more dangerous than lava.

When the press reached Subagan a few days later, thick clouds of steam were still rising over the village and howling dogs roamed the streets. 'Most of it is buried under six to 10 feet of black mud that fills its houses up to the roof and makes its roads unrecognizable,' the *Sydney Morning Herald* reported on 25 March. 'Thousands of boulders, some the size of cars litter the surface of the mud, next to crumbled walls, splintered telephone poles.'

Eight of the dead were found on the roof of the village's small mosque, one of them with minor burns but the others lacking any outward sign of physical injury.[1] On the nearby coast, crowds of women were described as standing, naked to the waist, knee deep in the ocean, 'praying to appease their gods and goddesses'.

With the picture of the devastated street on the seat next to me – it was a photocopy from microfilm and hence somewhat blurred – I drove into Subagan at that time of day when most people were either sleeping off the heat or sitting in deep shadow playing cards or chequers and listening to the radio. It looked pretty much like any other Balinese village that happened to be on a major road. There was a main street lined with single-storey shops and whitewashed bungalows, clouds of red and orange bougainvillea hanging over their garden walls. A motorcycle repair shop opened onto the pavement like a grease-blackened grotto. I looked closely at the shops and houses, comparing them to those in the picture, but none of them seemed to match, so I left the main street and drove slowly along a smaller road that led up a gentle slope behind the centre of the village. Here there were fewer shops and the houses were spaced further apart, with rice fields visible immediately behind them. A little further along the rice fields began to appear between the houses and I realized I had left the village behind. I stopped the car and got out where the sealed road became a dirt track that petered out at the edge of a steeper, bush-covered slope.

Large grey rocks in the fields to my left and right were surrounded by an expanse of luminous green rice growing in wide, low terraces. The neatly trimmed earth embankment of one of the terraces curved around an obtruding boulder and I wondered which had been there first – the terrace or the boulder. If the *lahar* had passed this way, as it surely had, then the terraces would have been rebuilt in its aftermath, the local people having no choice but to live with those rocks that were too big to shift. There were several boulders in the fields themselves, protruding above the surface of a rice crop that was like a green fluid that made islands out of anything it could not submerge. The level surface of the ripening rice was broken here and there, suggesting there were smaller boulders present but out of sight. The minutes passed while I stood interrogating the fields, enveloped by the heat. The trouble with the heat was not that it distracted you from your purpose but that, like a narcotic, it could all too easily lead you to fasten onto some insignificant detail while the afternoon slipped quietly away. Some women were working in one of the fields, their faces unreadable in the shade of their conical straw hats. They were looking over at me and laughing. One of them waved at me, or I thought she did, so I waved back.

Back down the road I stopped and looked over the low wall of an empty schoolyard. Projecting from the smooth dirt surface were several boulders too big to have been cleared away when the school was built or rebuilt following the eruption. Now that I had my eye in I began to notice rocks and boulders everywhere. Freshly washed clothes had been spread to dry over some big ones that lay in an uncultivated field of dry grass. Back down near the main road I stopped outside the bungalow-style office of some government department that was set in a carefully tended compound where the tops of boulders broke the surface of a smooth lawn, like features in a Japanese garden.

Upon returning to Subagan, the survivors of the eruption had begun life again and gradually restored the landscape to what it had been. The event that had effaced their neighbourhood was mostly effaced in its turn, but the continued presence of the partly submerged boulders illustrated the limits of this process of undoing (Figure 3.2). Unmovable and impractical to break up, they merged into the restored landscape. Rice crops and lawns flowed around them; garments that people wore on their bodies were laid out to dry on the surface of rocks that thirty years before may have crushed other bodies. I wondered if the boulders were now such a familiar presence that people no longer reflected on how they got there.

'Something confusing?' A young man was standing behind me, looking helpful.

'No,' I replied, nodding my thanks, 'I'm just looking.' Then, on the spur of the moment, I thought of showing him my photocopied photograph of the ruined street. Placing it on the bonnet of the car, I explained that it had been taken after the 1963 eruption, indicating with my eyes the direction of the mountain that was invisible from where we stood. For a few moments he gazed with interest at the picture of the devastation as motor scooters

Figure 3.2 Early to mid-twentieth-century shops in the vicinity of Jasri
Photograph by Denis Byrne

puttered past us and high school students in uniforms walked by in twos and threes.

'Where?' he asked.

I was momentarily thrown by his solicitous and seemingly sympathetic tone. Then I realized he had not understood that this picture was of Subagan, the village in which we were now standing.

'Here,' I replied. 'Subagan!'

'Ya, Subaaagan?' He picked up the picture now and looked at it more closely, but he shook his head. 'I think not Subagan,' he said, handing it back to me as if to rid himself of it. But I wasn't to be put off quite that easily.

'See, here, it says "Subagan",' I said, picking up the picture and pointing to the caption, suddenly conscious of the faith I place in the printed word.

He glanced sceptically in the direction of my finger but just then a small bus coming down the hill slowed to a stop in front of us and he quickly apologized and hurried off to board it, leaving me holding the picture, as if the eruption were my problem rather than his.

As I began driving slowly back along the main road in the direction of Candi Dasa I looked at the façades and roofs of the buildings more carefully. The tiled roofs all looked comparable in shape to those in the photograph and the black stains on the tiles, left by wet-season fungal growth, gave all of them an appearance of age. But that could happen almost overnight in this climate. Two of the shops in the photograph had façades in the *kantor* or 'office' style favoured by the Dutch administration in the first half of the twentieth century, but I could see nothing in front of me that quite matched them. I stopped the car and walked back along the street, pausing every few metres to take a close look at the buildings on the opposite side, trying to distinguish recent additions and renovations from older elements. In March 1963 an Australian Associated Press correspondent had written, 'Today Subagan is a dead town. Only vultures, and a few dogs and chickens and a number of looters are interested in it. Subagan is about as thoroughly destroyed as a town could be.'[2] And yet the present village looked as if it had always been there. It was a picture of solid and uninterrupted continuity.

Apart from the boulders I had just been observing, the 1963 eruption, as a historical event, had little or no visibility in today's terrain, at least not to my eyes. The landscape seemed to have grown over the evidence, like skin growing back over a graze, almost as though it were colluding in a form of censorship, encouraging a certain illusion. Was that what a cultural landscape was? A kind of template that simply replicates itself to cover any gaps?

I wondered how older people remembered the landscape that had been buried and semi-obliterated. Surely, even today, some must still be carrying a picture of it in their minds. I imagined a person of sixty in the fields, sowing rice seedlings, standing up to straighten their back and looking over at a hill slope covered in lush bush. Did they see the neatly trimmed rice terraces that had been buried there – terraces where they had worked all through their youth? Fields where, after school, they had sat under a palm-frond shelter, scaring the birds away from the ripened rice by pulling strings festooned with tin cans that were strung out over it. When they glanced in passing at a vacant piece of land, did they sometimes see an after-image of the house that had been there and the people who had lived in it – people who numbered among the lost? Or standing in the compound of one of the village's temples during a ceremony, did they sometimes visualize the old temple that had stood on the same spot and whose remains were buried below their feet?

A few days later, further towards the eastern end of the island, I followed a road that led to the upper slopes of Gunung Agung, winding up through a series of villages of decreasing size until the stony ground gave way to a lava field, at which point the road ceased. I had been assured by the local person who accompanied me that this had been fertile farmland before the eruption. Now it was beginning to be colonized by thinly scattered, tough, low shrubs and a resilient variety of grass. The place was desolate and dry. Some kind of shrub or tree was being grown from cuttings that had been stuck in the ground at regular intervals in a grid pattern. At first I failed to notice the grid: I just saw two or three of the sprouting sticks; then I saw more of them and the pattern snapped into focus. A solitary old man was gathering regular-sized rocks and placing them in neat piles. This was not for himself. It couldn't be. Surely he would be long dead before this austere terrain was productive again. For his descendants, then, or for the sake of something that was out of my grasp. For a moment I was appalled by the time span of what he was involved in.

Back on the streets of Subagan, meanwhile, my problem lay in deciding what in the streetscape dated from before the eruption and what from after. I knew it must have been possible to assign approximate dates to the structures on stylistic grounds, but, frustratingly, while I knew something of Balinese temple architecture, I was not sufficiently informed about modern Balinese secular architecture to make such discriminations.

When I left Subagan half an hour later and passed through Jasri, a village a few kilometres down the highway towards Candi Dasa, my attention was caught by the rough stone walls standing shoulder high in front of most of the houses. Was this where the smaller rocks deposited by the eruption had ended up? In the gardens of some of the houses there were ferro-cement imitation rockeries that took the form of small cliffs, sometimes with waterfalls and often with niches for growing ferns and trailing plants. These rockeries featured sham boulders, cleverly shaped and finished in cement tinted charcoal grey. You might have thought there were sufficient rocks in this landscape without the need to fabricate more.

Back in Candi Dasa at the small bungalow hotel I had chosen, situated at the end of the village, I sat on the verandah and looked along the curve of the shoreline. There was enough light left to see a few people sitting on the small aprons of white sand that some of the hotels had trucked in after the sea had removed most of the original beach. Closer to me, on a surviving remnant of that old beach, a fisherman was unloading a double outrigger canoe. A slight figure, burned dark brown by the sun, he walked up the slope of the beach towards the coconut palms, carrying his folded net over one shoulder and a string of small silver fish in his hand – a pathetically small reward for a day spent at sea. Like the youths you saw trolling along Kuta Beach day in, day out, hawking fake Rolexes that were next to impossible to sell, it was an insight into the economics of poverty. I was still sitting there when night fell. The boy who was watering the garden with a low-pressure hose, working his way down the strip of shrubs in front of the cottages, had finished and left. I thought about my own day and the little I had to show for it, but mostly I just looked at the waves breaking over what was left of the reef and at the much smaller order of waves coming in this side of the reef. They moved across the lagoon in a regular pattern and came within a hand's breadth of covering the top of a wooden pole that had been driven into the seabed.

I was woken in the night by the noise of a motorbike somewhere nearby. Its engine died and was followed by the sound of footfalls scrunching along the gravel path. I lay on the hard mattress, conscious of the stickiness of my skin and the coarseness of the sheets, and when I knew I would not sleep again for a while I got up and went back to the seat outside. The moon illuminated the scalloped edges of the clouds as it passed behind them. I could sense no movement in the air but, even so, salt-laden moisture was rising off the breaking waves and settling in a film on the tiles of the verandah, leaving them tacky underfoot. It settled invisibly on my bare arms, too. After a while I heard a door close and a young Balinese guy with the Rasta look of the Candi Dasa gigolos came back down the path, nodded to me as he passed, and rode away on a motorbike.

For want of anything better to do, I picked up my folder of 1963 press clippings and, with my back against the cool masonry wall, started reading through the stories that shared space on the same pages as the reports of the Gunung Agung eruption. In Singapore in March 1963 they were talking about water rationing and the upcoming meeting in Washington between the Malayan deputy prime minister and John Kennedy. The Profumo affair was breaking news in London, while in Tokyo Crown Princess Michiko had reportedly had an abortion. France had just detonated a nuclear device under the Sahara Desert; students in Algiers were protesting against the test and the United States expressed a wish that the French would carry out their experiments in the Pacific instead. Meanwhile, Queen Elizabeth II was making her way majestically down the coast of Western Australia in the royal yacht, *Britannia*. Like me, she may have lain awake in the night, listening to the sea and wondering if she was getting the most out of life. The *Singapore Times* reported that 75,000 students in Bandung, Java, had marched on the residence of the military governor, protesting against rising prices and demanding communist representation in Sukarno's cabinet.

Erosion

The following day I drove to Amlapura, the old seat of the rajas of Karangasem and only a kilometre or so east of Subagan. It was a quiet place of tree-lined streets and swept pavements whose commercial centre consisted of a single strip of two- and three-storey shop houses, mostly Chinese-owned. But it was the only place in the Regency of Karangasem that was larger than a village, and if it had gone undamaged during the eruption it had also been bypassed by the tourism and development boom.

I had been told they rented out rooms at the Puri Agung Karangasem, the old raja's palace, but when I parked the car on the street outside and climbed the steps to the first of a series of walled courtyards, this hardly seemed likely. The palace extended along the top of a ridge and the courtyards were terraced into its flanks. Just inside the monumental brick gateway to the first courtyard was a small table on which lay a ledger book and a block of admission tickets. There was nobody around so I took the steps up to the main courtyard, most of which was taken up by a carp pool, half covered with lotuses, in the centre of which was an island pavilion reached by a small causeway supported by arches. It was a delightful space but there was no sign or promise of hotel accommodation.

When I turned to retrace my steps down to the first courtyard, I was confronted with an unobstructed view of Gunung Agung's cone (see Figure 3.3), the outline of which was softened by the haze. I had missed it on the way in, too absorbed by what was in front of me to have bothered to turn around. Down below, the ticket seller was now sitting at his table. I learned over the next couple of days that the trickle of tourists who found their way to the palace often dried up for hours on end and during these spells the ticket seller, who introduced himself as Madé, could stretch out and nap in a small, open-

Figure 3.3 Gunung Agung seen from Puri Agung Karangasem, 1994
Photograph by Denis Byrne

sided pavilion nearby. Madé had several fighting cocks, lined up on the gravel in light rattan cages shaped like bell jars. They needed a lot of handling and pampering and moving in and out of the shade and this took up a lot of his time.

Madé led me off to see one of the rooms that was for rent, taking me up to the main courtyard, past the lotus pond, to a smaller gateway beyond which stretched a dirt path hemmed in by the outer walls of the succession of small courtyards opening off it and to which it provided access (Figure 3.4). Down this path we passed two of the courtyard gateways, which, similar to those found in temple compounds, were like ornate brick bookends placed close together. As always, there were a couple of brick steps on either side of the actual door that was set between the bookends, so you stepped up through the gate and then down into the courtyard, in a movement that marked the crossing of a liminal threshold. Since temples, houses and, obviously, the old palaces of the rajas consisted not of bulky single buildings but free-standing rooms and open pavilions distributed around the enclosed spaces of walled courtyards, the entry gateway took on the significance that a front door would have in a Western building. Even the simplest of gateways in Bali had winged doors that opened inwards from the middle, so you always passed through the centre of the opening rather than somewhat to the left, as you would with a single-wing door.

Figure 3.4 Pathway at Puri Agung Karangasem, 1994
Photograph by Denis Byrne

Nothing seemed to move in the courtyards that we passed except the leaves of the trees. The bare, weatherworn wood of the doors and the patches of orange brickwork exposed through layers of exfoliating plaster evoked the atmosphere of gradual decline that permeated the place as a whole. Madé showed me into the courtyard that contained the guest-room pavilions, none of which was occupied. Copying him, I slipped off my sandals, stepped onto the paved verandah of one of the rooms and passed into the dim space of its interior. It was hotel accommodation at its most basic but entirely in keeping with the mood of the palace.

Madé departed and I strolled around, looking at the plants, but was then drawn back to the dirt pathway we had just come along. The dirt seemed to have been swept wall-to-wall that morning with a *sapu lidi* – one of those bunches of stiff straw in universal use on the island. Only our footsteps violated the path's velvet surface. The sweeper's brushstrokes were even visible around the trunks of the hibiscus and the stems of the smaller shrubs planted next to the walls on either side. Looking more closely, I could see the fresh tracks of lizards.

The white plastered surfaces of these walls were cracked and discoloured by grey and green mould which had spread in patterns that mapped underlying inconsistencies in the layer of render. The gateways were of unplastered orange brick, with the bricks about half the size of standard Western bricks. They were laid with almost no mortar so, from a distance, the joins were almost imperceptible (Figure 3.5). One of the characteristic aspects of Balinese architecture comes from the low-fired nature of these unglazed bricks. They erode relatively quickly, although some disintegrate faster than others to produce rockshelter-like rectangular concavities in the otherwise smooth orange surfaces of walls and pillars. Three or four such 'soft' bricks were eroding on the outside face of the gateway to my courtyard, spilling a powder of bright orange particles onto the ground a metre or so below, forming small half-cones of detritus at the base of the wall. I squatted down to get a better look, thinking I might see some particles fall.

There was traffic noise in the distance and the tinny sound of a radio wafted over the wall from a nearby courtyard. Level with my head was the cavity formed by the eroding bricks. Here the flat surface formed by the underlying course of uneroded bricks provided a ledge upon which a small talus slope of brick dust was accumulating before dropping to the ground. My eyes moved from the talus to the tiny pile of particles on the ground below and I was aware for a moment of how this whole complex of buildings was gradually breaking down and that this was somehow intentional. Buildings all over the island were decaying in this manner – the hourglass of the falling brick dust calibrating the rate of attrition. Rarely was anything done to interfere with this process of decay. People on the island lived with a level of decay that, though relatively rapid, was entirely sustainable, matched as it was by their ability to rebuild.

The pathway under my feet would be swept every morning in one of the first acts of the day. The broom would gather up the brick dust and plaster

Figure 3.5 Eroding brickwork of a gateway at Puri Agung Karangasem, 2017
Photograph by Denis Byrne

fragments from the exfoliating wall along with fallen blossoms and cigarette butts. Like the Thais, the Burmese, the Javanese and others of the region, the Balinese were meticulous sweepers of the space around built structures, leaving the structures themselves to get on with the business of deterioration.

While I was absorbed in the erosion of the brick wall, a part of my mind was still on the Gunung Agung eruption. I had a vague sense of there being a link between the two, but precisely what it might have been eluded me. I also carried with me residue of previous years I had spent looking at decaying Buddhist temples and stupas in Thailand as part of my doctoral fieldwork (Byrne 1993). The tropical climate of the region took a toll on the brick and stucco masonry of these structures, and termites consumed the woodwork. But the constancy of attrition and decay was apprehended by people not so much in terms of loss as of opportunity eventually to gain merit by restoration and rebuilding (Byrne 1995; on Buddhist merit in Thailand, see McDaniel 2014; Tambiah 1984; see Figure 3.6). I had seen old stupas that had split apart in the course of earthquakes, exposing within themselves the remains of earlier stupas whose decayed state had created opportunies for merit-making by those who had encased them within the masonry of the new structures (Byrne 2015: 86–87). The life of each stupa is one of disintegration and accumulation. But for every stupa or temple restored there were others that were simply left to decay, their ruins subsiding into the ground or colonized by jungle.

In many respects, heritage conservation is a project of containment. The aim is to contain the fabric of old buildings within the body of old structures rather

Figure 3.6 Fishpond in a courtyard at Puri Agung Karangasem, 1994
Photograph by Denis Byrne

than allow it to be removed by looters, recycled as building material by local farmers, carried off in particles by termites and other insects, or collected as numinous talismans by religious believers. Though radically different in nature, these are all forms of dispersion that undo the process of gathering or assembling that brings an object such as a wall into being in the first place (clay assembled into bricks, for instance). Against the numerous natural agents of dispersion, the conservator seeks to stabilize the building and its material fabric, shoring it up, holding it in place. Stabilizing things in this way may create the illusion that they are static and timeless by nature, rather than being endowed with a 'socially produced durability' (Buchli 2002: 15; see also Lucas 2013 on the tension between the aesthetics of ruin and the modern urge to sanitize and preserve). By contrast, when pious Buddhist believers in Thailand eventually 'restored' their temples and stupas, they made every effort to display the scale of their interventions, which often resulted in structures that bore little resemblance to the originals. By the same token, prior to restoration, they made no effort to conceal the dilapidated state of an old structure.

Conclusion: travel in retrospect

I was no disinterested observer on that morning in July 1994 when I squatted on the pathway at the Karangasem palace in the hope of catching particles of eroded brick in the act of falling. In retrospect, I see a connection between the eroding wall and the 1963 Gunung Agung eruption, one that had eluded me in 1994. The falling brick dust carried an implicit allusion to the volcano's ash shower (and *lahar*), each representing an undoing by natural forces of the Balinese built environment and cultural landscape. What propelled me on my journey in 1994 was a desire to see how people *lived with* that unravelling, that loss. I keep going back, in my mind, to the figure of the old man gathering igneous rocks on the lava field, gradually putting the former agricultural field back together or working to rebuild a new version of it. He could be described as acting out his geosubjectivity, in the sense that Kathryn Yusoff (2015) develops that term.

Yusoff's (2015, 2016) concern with humanity's intense relationality with coal is part of an effort to break down the artificial boundary between the human and the geologic at a point in human history where it has become a matter of urgency that we realize we live *inside* the earth system rather than enjoying the position of being external observers of it, with all the implications of superiority, mastery, and 'human exceptionalism' (Braidotti 2013: 36) which that carries. Taking up Helene Cixous's notion of 'blood knowledge', Yusoff (2015: 206) offers a way of seeing humans as sharing a 'line of descent' with coal, in terms of which our desire for coal and the pleasures it affords and our embodiment of coal through inhaling carbon emissions are but two of the ways we live with and in a Carboniferous material temporality. The project of reconstituting the lava field as farmland – an ambition I project onto the old man encountered on the upper slopes of Gunung Agung – sees him working in concert with the weather, the bacteria and invertebrates and plants that are colonizing the lava, and with the igneous rock's chemical properties and their soil-forming capacities. I construe him as forming an alliance with these more-than-human actors, each of which has its own trajectories and 'ambitions' and its own world-making alliances. As Anna Tsing (2015: 22) puts it, 'each organism changes everyone's world. Bacteria make our oxygen atmosphere, and plants help maintain it. Plants live on land because fungi made soil by digesting rocks.' We can thus understand that to live under a volcano is to live in/with the geologic, not just to live subject to geological events.

The things we encounter in our travels often have a certain vividness, much of which no doubt stems from our abnormal circumstances – being in a strange place, in a different time zone, encountering strange people. But these encounters often stay with us. They have an unfinished quality, which is something that may be true in a particular way for academic travellers: we periodically rehearse these encounters, finding new meaning in them in the context of new projects we undertake. We are never quite done with our travels and this has certainly been true of my 1994 experience on the slopes of Gunung Agung.

I felt the truth of this most recently in circumstances which, appropriately enough, saw me standing next to another wall, watching it erode. The wall in question was the seawall at Elizabeth Bay, on the south side of Sydney Harbour (Byrne 2017). Built in the 1880s out of large sandstone blocks, the wall contains the infill deposit used to reclaim the inner extremity of the small bay in order to create a harbourside park. It was also designed to protect the reclamation from the harbour waters. In the course of a study of the history and archaeology of coastal reclamations, one afternoon in 2015 I found myself standing at low tide on the narrow strip of beach at the foot of this wall – a wall that previously I had construed as a static barrier to the sea. Now, however, I noticed how quickly the stone blocks were eroding. Like the bricks in the wall at Karangasem, the sandstone blocks were incrementally giving up their substance, which, in their case, meant that the quartz grains that had been sedimented into rock 200 to 400 million years ago were falling to join the beach sand at the foot of the wall. While the disaggregation of the bricks in the Puri Agung wall resulted in an inconsequential widening of the lane by a few millimetres each year (as the bricks shrank back), the seawall's disaggregation meant the sea was able to advance a few millimetres each year into the space of the reclamation, back towards the beach that in the period prior to the 1880s its waves had been free to run up.

What began for me in Karangasem was an awareness of temporalities that I had previously been blind to: the particular temporalities of the old man's project of restoration and of the brick wall's dissolution. I cannot say with any certainty that this 'opening' set me up, decades later, to tune in to the temporality of the seawall at Elizabeth Bay. But I suspect that travel experiences are always simultaneously behind us and present in palimpsest form here at the surface of our present endeavours.

Notes

1 Eastern Bali has a small but significant population of Muslims whose presence there dates from the period prior to the mid-nineteenth century when the predominantly Muslim island of Lombok was ruled by the rajas of Karangasem.
2 Published in the *Sydney Morning Herald*, 26 March 1963.

References

Booth, W.P. (1963) 'Disaster in Paradise', *National Geographic*, 123(3): 436–447.

Braidotti, R. (2013) *The Posthuman*. Cambridge: Polity.

Buchli, V. (2002) 'Introduction'. In: V. Buchli (ed.) *The Material Culture Reader*. Oxford: Berg Publishers, pp. 1–22.

Byrne, D. (1993) 'The Past of Others: Archaeological Heritage Management in Thailand and Australia'. Ph.D. thesis, Australian National University, Canberra.

Byrne, D. (1995) 'Buddhist Stupa and Thai Social Practice', *World Archaeology*, 27(2): 266–281.

Byrne, D. (2015) *Counterheritage: Critical Perspectives on Heritage Conservation in Asia*. New York: Routledge.

Byrne, D. (2017) 'Remembering the Elizabeth Bay Reclamation and the Holocene Sunset in Sydney Harbour', *Environmental Humanities*, 9(1): 40–59.

Lucas, G. (2013) 'Ruins'. In P. Graves-Brown, R. Harrison and A. Puccini (eds) *The Oxford Handbook of the Archaeology of the Contemporary World*. Oxford: Oxford University Press, pp. 192–203.

McDaniel, J. (2014) *The Lovelorn Ghost and the Magic Monk: Practicing Buddhism in Modern Thailand.* New York: Columbia University Press.

Tambiah, S.J. (1984) *The Buddhist Saints of the Forest and the Cult of Amulets.* Cambridge: Cambridge University Press.

Tsing, A.L. (2015) *The Mushroom at the End of the World: On the Possibility of Life in Capitalist Ruins.* Princeton, NJ: Princeton University Press.

Vickers, A. (1989) *Bali: A Paradise Created.* Ringwood, Victoria: Penguin.

Yusoff, K. (2015) 'Queer Coal: Genealogies in/of the Blood', *PhiloSOPHIA*, 5(2): 203–229.

Yusoff, K. (2016) 'Anthropogenesis: Origins and Endings in the Anthropocene', *Theory, Culture and Society*, 33(2): 3–28.

4 Facing a new frontier

Defining nature and culture within dinosaur parks in Europe

Ross J. Wilson

Introduction

This chapter builds upon the recent work on the phenomenological experience of heritage sites to examine how modern dinosaur parks in Europe can be used to understand how humans and non-humans interact in the age of the Anthropocene. Rather than assume that phenomenology entails a solipsistic assessment, this chapter utilizes the wider application of the theoretical approach which emphasizes a being-in-the-world with others (after Gurwitsch 1977). Through engagement with the models of the extinct creatures, Latour's (2004) principle of intertwining or enmeshing with nature can be explored, which offers an alternative vision of a 'social environment' beyond the homocentric. Therefore, dinosaur parks represent a significant resource to rethink humanity's place within the world. This stands in contrast to the absence of studies regarding this highly prominent aspect of natural heritage (see Wilson 2017). This neglect may derive from the status of these parks as commercial operations or children's entertainment. However, these sites have been prominent parts of wider society's engagement with natural history for over a century. Beginning with the creation of 'lifesize' replicas of these extinct animals in the late Victorian period, this natural history has exercised the emotions of audiences as a means of entertainment but also a point of reflection. The 'terrible lizards', as they were first defined, constituted objects of fear not just through their dimensions and presumed ferocity but because their disappearance was regarded as a fate that might also befall humanity (Marshall 2007). With humanity's place on Earth under threat through climate change, these sites appear to be ever more relevant for the modern world (after Mitchell 1998). Over the last three decades the number of dinosaur parks has increased across European countries as the portrayal of these animals in film, television and literature has enhanced public interest. As we enter the Anthropocene, the survival of humans will require an alteration in our engagement with the environment; it is within the dinosaur parks that we can begin to consider 'being-with-others'.

Dinosaur parks and popular culture

The array of dinosaur parks that exist across Europe today is a legacy of the cultural, social and intellectual shifts over the last two hundred years. With the discovery of large fossilized bones during the early nineteenth century in Europe and North America, scientific and public fascination with these ancient creatures greatly increased (see Moody *et al.* 2010). Displays and public lectures on the scale, appearance and possible causes of the disappearance of the animals were commonplace. As new specimens were uncovered, the biologist and early palaeontologist Richard Owen (1804–1892) capitalized on this interest by coining a new term to classify the discoveries (Owen 1841). 'Dinosauria' was formed by combining two ancient Greek words – *dinos* ('terrible') and *sauros* ('lizard'). Defining and characterizing these specimens as a specific order of animals served to heighten their appeal to the public. The excitement and fear generated by the size of the fossils was thereby reflected in their new name, which added a sense of mystery and exoticism to these long-extinct creatures (Jaffe 2000). Illustrations and models of the 'antediluvian beasts' served to cement the presence of 'saurians' within the popular imagination (Rudwick 1995). Indeed, such was the clamour for the latest discoveries in the field that the world's first dinosaur park was opened in Crystal Palace, south London, in 1854 (Figure 4.1). Under the direction of Owen, the artist Benjamin Waterhouse Hawkins (1807–1894) created a series of 'realistic' sculptures of fifteen extinct animals (Doyle 2008: 197–198). Although there were renderings of prehistoric mammals, the most prominent sculptures were those of *Ichthyosaurs* and *Plesiosaurs*, for which the arrangement was named the 'Dinosaur Court' (see Owen 1854).

The confrontation with the dinosaurs established a distinct relationship between humans and nature (Secord 2004). Significantly, this engagement focused on the presence of the extinct animals as a demonstration of their inefficiency and inadaptability to change. As such, the Dinosaur Court acted as a point of reflection for nineteenth-century audiences to regard the advances of their own era (see Society for Promoting Christian Knowledge 1852). The juxtaposition of the presence of ancient history within the modern era that was captured in this display became a popular trope within novels throughout the late nineteenth century and featured in prominent examples of early film (see Conan Doyle 1912). The 'playful' interchange between the frontiers of past and present alongside public and academic proved profitable for authors and artists as popular interest in dinosaurs was fuelled by the identification of new discoveries such as the *Stegosaurus* (1877), the *Diplodocus* (1878) and the *Triceratops* (1889) in North America (see Brinkman 2010; Wallace 2000). This fascination continued with models of dinosaurs displayed at World's Fairs in the early twentieth century, whilst museums of natural history also began to use sculptures to educate and inform visitors (after Luckhurst 2012) from the 1930s. Dinosaur sculptures were used within a commercial context across the

Figure 4.1 Dinosaur Court, Crystal Palace Park, London
Photograph by Ross J. Wilson

United States, drawing upon the age or size of the extinct animals to raise awareness of products and services in the present (see Merkl 2015).

The use of dinosaurs within popular culture can be observed with Sinclair Gas and Oil's display at the World's Fair in New York in 1964. Nine life-size replicas, including a *Stegosaurus, Brontosaurus, Triceratops* and *Tyrannosaurus Rex*, were placed in an enclosure designed to mimic a Mesozoic environment. The popularity of the models was such that they were distributed to natural history museums across the United States after the conclusion of the fair (Samuel 2007: 174). The excitement and interest generated by these displays also encouraged the proliferation of commercial dinosaur parks within North America (see Costa 1994). By the 1960s, the trend for dinosaur models as educational or touristic sites was being developed across Europe, too, as parks, gardens, zoos and theme parks began to use both statues and animatronic models to attract and entertain visitors. With a discernible 'dinosaur boom' in the 1970s, when public interest rose again, and then an array of films and television programmes, such as *Jurassic Park* (1994) and *Walking with Dinosaurs* (1999), these animals have attained even wider prominence, and museum displays and attractions have increased (see Bakker 1975). Indeed, it seems that across sites of varying size in Europe, the dinosaurs have been resurrected.

Natural history and emotional engagement: heritage at the frontier

Despite the development of these places of natural history, the study of this heritage has been largely absent from wider scholarship (Wilson 2017). Whilst numerous studies have assessed issues of conservation and protection for sites of natural history, critical engagement with how this heritage is used, valued and represented within contemporary society has not been similarly explored (after Harrison 2015). However, as pressures on the environment and climatic change begin to alter societies across the globe, the relationships that define 'nature' and 'culture' will be placed under greater examination (Harvey and Perry 2015). Heritage becomes an essential discipline in this context as it provides a means by which those associations between people and the environment can be redefined (Solli *et al.* 2011). This can be achieved by altering the perception of humanity as removed or distanced from nature, which is how it was defined within Enlightenment culture within Europe from the sixteenth century (Ingold 2000). The development of scientific, logical and rational procedure fostered the perception of humans beyond the involvement of the natural world (after Latour 1993, 2004). Such disassociation from a wider ecological context has been assessed as a contributing conceptual factor that enabled the rapid industrialization of Western society from the nineteenth century to the present (Morton 2007). This has laid the foundations for a society that regards pollution and environmental damage as acceptable consequences of progress and therefore issues to be managed rather than eliminated. To address such damaging and limiting perspectives, 'natural heritage' can recast humanity's relationship with the wider world by assessing our social engagement with the environment.

The study of how visitors physically and emotionally interact within heritage sites has developed considerably over the past ten years, with scholarship focused on how places and objects exercise and engage with individuals to elicit feelings and exercise sentiments (Smith 2006; Smith and Campbell 2015; Tolia-Kelly *et al.* 2017). It is in this regard that the use of phenomenology within Heritage Studies has been drawn upon to assess how places structure experience (after Masberg and Silverman 1996). Phenomenology represents the development of a range of approaches after its intitial definition as a study of consciousness by Husserl (1964). Developing from an awareness of an individual, phenomenology was expanded by Heidegger (1962) and Gadamer (1975) to explore hermeneutical phenemonology. This concept addressed how an understanding of the world was obtained through our awareness and engagement with our wider environment (Heidegger 1962: 78). This necessitated a movement towards recognizing the social in the formation of the individual; that we engage in an act of interpretation with others (see Schütz 1967). Gurwitsch (1977) argued that a phenomenological analysis required an awareness of how our 'being-in-the-world' was dependent on encountering

and understanding other beings. The use of the 'social' in the creation of meaning offers a process of expanding our understanding of the term to encompass both humans and non-humans (Latour 2004). Therefore, a hermeneutical phenonology can explore how meaning is created with the wider environment rather than imposed by humans upon the world (Carel and Meacham 2013).

Phenomenology can serve to bridge the conceptual divide between 'nature' and 'culture', as to be a 'being-in-the-world' always necessitates an engagement between the human and the non-human (Bannon 2014). This challenges studies of political ecology which assess the relationship between people and nature as a process of reification, with notions of the 'environment' accorded primacy without regarding its construction within the realm of the 'social' (after Latour 2004). Through this approach, a phenomenological examination of natural heritage can serve to broaden our frontiers and examine how humans and non-humans alike exist within the world. This builds upon the recent work within Heritage Studies that utilizes phenenomology to assess how visits are intersected with varying factors that structure the experience of sites and landscapes (Cohen 1996; Uriely 2005). Dinosaur parks can be examined on this basis as they represent a key means of engaging with natural heritage and environmental history and place humans within a direct context of non-humans.

To assess how dinosaur parks provide this engagement with the 'social' experience, thirty sites across Europe were studied.[1] These range from purpose-built dinosaur parks located near the discovery of significant fossil finds to dinosaur models displayed as part of a commercial venture, safari park or zoological gardens. The survey documented exhibits, maps, guidebooks and websites to focus on the key elements of each park:

- language – the discourses representing the site;
- structure – the spatial organization of the site; and
- place – the objects and exhibits that form the perception of the site.

From large commercial enterprises to small-scale family businesses, these dinosaur parks have been formed through different processes but they all possess common traits in their representation of how the ancient past is brought into our own era. The presence of these animals within the contemporary era is not cast as an unsettling intrusion that threatens the modern age, nor do the parks attempt a straightforward sense of continuity between past and present (after Noble 2016). Rather, the dinosaur parks place humans within a context where their dominance in the environment is challenged and we are required to consider ourselves in the context of the non-human (after Brown and Toadvine 2003).

This engagement can be overlooked within the existing paradigms of a consumerist model within Heritage Studies (see Wilson 2014). In this assessment, dinosaur parks could be cast as nothing more than a parade of the hyperreal (Eco 1986). The construction of models of animals from the Triassic (252 MYA–201 MYA), Jurassic (201 MYA–145 MYA) and Cretaceous eras (145 MYA–65 MYA) could be considered to affect the 'desert of the real' (after Baudrillard 1994). Reflecting a late capitalist desire to consume, offering a spectacle of the past as a commodity rather than engaging society with reality, dinosaur parks can be construed as symptomatic of the wider malaise within modernity (Lyotard 1984). However, such assessments would obscure the way in which these sites provide a setting for the engagement between humans and non-humans by concentrating on political alienation rather than public engagement. In considering the latter, it is the experiences and perceptions framed by the dinosaur parks that demonstrate a departure from established modes of being. It is within the coexistence of past and present exposed by these sites that an alternative vision of nature and culture can emerge (after Žižek 2014). By placing humans and non-humans together within these sites, the forces of modernity that divided 'nature' from 'culture' can be reconsidered (after Latour 1993). To be a being-in-the-world in these spaces is to encounter another understanding of the social beyond the human (after Latour 2004).

Presenting excitement and fear in the dinosaur park

Within the websites, promotional literature and signage of Europe's dinosaur parks, the language that defines these areas focuses upon a sense of excitement and discovery (after Duff 2010). It is in this process of the encounter that a sense of identity with 'the other' is formed (Gurwitsch 1977; Schütz 1967). This can be seen at the Dinosaurier Park, an open-air museum in Münchehagen, Lower Saxony, Germany. It is one of the largest dinosaur parks in Europe and opened in the 1980s after the discovery of hundreds of dinosaur tracks from the Cretaceous Era. The park provides trails through the area where visitors discover animatronic models of the extinct animals from a range of periods during the Mesozoic Era. This engagement is phrased within a discourse of 'the encounter'. Each individual visitor is encouraged to feel a sense of the presence of 'the other' as they come across each exhibit: 'In the open-air area, visitors can encounter not only all the famous dinosaurs such as *Stegosaurus, Triceratops, Brachiosaurus* and, of course, the *Tyrannosaurus rex* but also the most interesting extinct animals of evolutionary history' (Dinosaurier Park 2016b).

A similar mode of engagement can be seen at Saurier Park, near Bautzen in Eastern Saxony. This is Germany's other major dinosaur park, with several hundred specimens displayed throughout trails that are divided into specific eras (Saurier Park 2016). The site presents itself as a place of 'excitement, fun and adventure' for children and adults alike. However, it is the confrontation with the past that is presented as significant:

> The giants of Jurassic time like Brachiosaurus, Diplodocus or Barosaurus will impress you by their enormous size. Large predators like Allosaurus, Gorgosaurus and the famous Tyrannosaurus Rex let you feel their former danger. But also many smaller dinosaur species with interesting forms can be discovered and admired.
>
> (Saurier Park 2016)

It is the affect of meeting the extinct creatures at these sites that is emphasized in this discourse. Such engagements do not require the individual to be elevated from the everyday to experience. Rather, it is their everyday self that provides a means of encountering something beyond the homocentric (Brown 2003: 6). As such, Dinosaurier Park, whilst focusing on a scientific mission to communicate the history of evolution, uses an approach based on 'exciting moments of experience with practical work' to reveal the presence of this ancient past as a lived being-in-the-world (Dinosaurier Park 2016a). It is this engagement with others that establishes a wide sense of the social environment (after Schütz and Luckmann 1973: 3–4).

The same process is explored in the array of dinosaur parks within the Czech Republic, which have been developed as private enterprises within the past decade. Indeed, the same leisure and entertainment company currently operates sites in the Czech Republic, Slovakia and Spain. Each of these locations features models and animatronic dinosaurs. The first to open in the Czech Republic was the Prague venue in 2011, which is situated on the rooftop of a shopping mall as a means to attract shoppers. Its displays of specimens from different periods are used to emphasize the sense of excitement and wonder as its entrance gates are modelled to appear as if it were a safari, with dinosaur models looming over the shops below (DinoPark 2016). The visual, material and textual prompts within these sites emphasize the encounter with 'the other' as part of a social environment that asks visitors to question the wider status of *Homo sapiens* within contemporary society (after James 2016: 43–44). Visitors are provided with a setting where humanity's role as the dominant life on Earth is questioned and sometimes placed under a playfully implied threat.

This sense of danger is also explored in Zator, Małopolska, southern Poland, where DinoZatorland opened in 2009. The site provides leisure attractions as well as a range of animatronic models of dinosaurs from the Triassic, Jurassic and Cretaceous periods (DinoZatorland 2016). However, it is the presentation of the site as one in which fear and the unknown operate to transport the individual to another age whilst never leaving their own time that serves to alter a sense of being within the park:

> The biggest attraction is our nature trail in the old forest, which is rich not only in 'animated' prehistoric animals, but also in beautiful centuries-old trees. It is here where you can admire the world's largest mobile T-Rex! This walk will allow you to move back through time for hundreds

> of millions of years. You will also be able to listen to the terrifying sounds of dinosaurs.
>
> (DinoZatorland 2016)

The same discourse of discovery is utilized at Givskud Zoo, Vejle, Denmark, with the development in 2015 of a dinosaur park to complement the display of animals within the venue's zoological gardens and safari park. The display of forty models has been highly successful in attracting high visitor numbers, but it is the presentation of these extinct creatures as points of present-day concern that defines the site:

> Imagine travelling millions of years back into time. To a time when humans did not even exist on Earth. To a time when the climate was hot and the land mostly consisted of coniferous forests and widespread swamps. Denmark was completely covered by water. This was the age of dinosaurs … you have the chance to see them life-size … The dinosaurs ruled the earth for 150 million years. So far, man [*sic*] has only been around for approx. one million years.
>
> (Givskud Zoo 2016)

The dinosaur park becomes a means to address the relationship between humanity and the wider world (after James 2009: 18). Through these modes of representation, the models are used to destabilize the established relationships between humans and the environment and consider new frontiers (Bannon 2016: 53–54). This is most clearly represented in the playful use of signage within several of the dinosaur parks that emphasize the 'dangerous' nature of the models. Notices advising visitors to stay on the trails and not to anger the dinosaurs are common across European dinosaur parks. In West Midland Safari Park's 'Land of the Living Dinosaurs', which opened in 2015 as an addition to the well-established animal displays, around the animatronic models are signs that caution visitors that the attractions could threaten them, or vice-versa (Figure 4.2):

> Do Not Touch: They don't want your germs
> Keep to the Paths: If not, the dinosaurs may eat you and that might make them sick
>
> (West Midland Safari Park 2016)

The language that represents and frames the experience of the dinosaur parks to visitors focuses on unsettling the established notions of self, place and time within contemporary society. In such a setting, new forms of being-in-the-world are formed in relationship to others (Gurwitsch 1977). The sites require individuals and communities to consider their place in the environment in an alternative manner where humans are not dominant and share the world with non-humans (after Latour 2004).

Figure 4.2 Sign at the West Midland Safari Park
Photograph by Ross J. Wilson

Organization and structure of the dinosaur parks

The dinosaur parks of Europe might vary in size and differ in their institutional arrangement but the structure of the parks follows a similar pattern and layout. Each of the parks is built around trails and pathways that guide the visitor through scenes and displays of dinosaur specimens that emphasize connection and reflection rather than chronological separation. As a being-in-the-world within these sites, humans are required to regard their place in the environment as temporal and as part of a far broader society of the human

and the non-human (after Greaves 2016: 31–32). In Bałtów, Świętokrzyskie Voivodeship, Poland, JuraPark was created in 2004 near the site of fossil discoveries. Covering an area of three hectares, it is one of the largest tourist sites in Poland. Two other dinosaur parks run by the same company are located at Krasiejow and Solec Kujawski, also in Poland. The Bałtów park is dominated by its dinosaur models, although a leisure complex, theme parks, restaurants and spas also attract visitors (JuraPark 2016). However, all three parks are experienced by paths that are arranged amongst the models and dioramas of dinosaurs from the Mesozoic Era. These paths provide points of discovery but they also enable a means of regarding the models as part of a social world as visitors interact with the creatures they encounter (after Schütz and Luckmann 1973: 15–16). The Préhisto Dino Parc, in Lacave en Quercy, Lot, southwest France, also provides a means through which this engagement between humans and non-humans can take place (Préhisto Dino Parc 2016). Opened in 2014, this park features a display of dinosaurs along a woodland trail. The specimens range from some of the earliest known sauropods to animals that inhabited Earth at the end of the Jurassic Era. Through its paths, Préhisto Dino Parc is orientated towards the social, where the lives of others beyond the self are experienced, engaged with and understood:

> Dinosaurs, these impressive creatures that have always amazed us and that occupied our planet 250 million years ago, could reach nearly 15 m high and weigh over 20 tons. During your visit in Préhisto Dino Park, you will be able to observe closer these ferocious predators ... You will discover the lifestyles of these giants through the ages, their habitats and their eating habits ... Dinosaurs will soon have no more secrets for you.
>
> (Préhisto Dino Parc 2016)

The park thus becomes a place where a new sense of being-with-others can be realized; where human existence is not inevitable but contingent and our place within the environment is regarded not as pre-eminent but as a component of a far wider society that encompasses the non-human (see Bannon 2014: 18–19). Whilst this epiphany might occur next to plastic and fibreglass models of creatures that have not lived for millions of years, the affect is still significant. In Prague's DinoPark, the models stand adjacent to the tower blocks of the modern city, forming a collocation of the prehistoric and the present. This arrangement enables a complex sense of being that transcends chronological periods and homocentric notions that define the Anthropocene (Head 2016: 39–40). As Mitchell (1998: 8–9) suggested, the figure of the dinosaur is perhaps the iconic symbol of the modern age, revealing a desire to dominate and control but fraught with potential destruction. Therefore, it is in this engagement at the dinosaur park that humanity can consider its place in relation to the wider world. The presence of a species that once proliferated

across the Earth places contemporary human society in a wider collective as it moves visitors beyond a recognition of a synchronic status of the world and into an acknowledgement that this place has been inhabited and is still inhabited by humans and non-humans alike (after Bannon 2014).

The transformation of identity and place that can occur within the dinosaur park can be observed within a number of sites in France. For instance, the Musée-Parc des Dinosaures in Mèze, Hérault, was developed in the last decade on a site of extensive fossil discoveries. Full-scale replicas of animals from throughout the Mesozoic Era are exhibited along paths within the woodland park. Alognside the trails are small exhibitions featuring fossils and data from the local area, which encourage engagement with the past and the present. The physical process of moving through the site places humans alongside non-humans, providing a point of reflection on the scale of humanity in comparison with the deep time of the Earth (after Sessions 1995). The paths engage visitors with an era and a perspective beyond their immediate environment to create a 'public meeting' between the present and the distant past (Musée-Parc des Dinosaures 2016). This 'public meeting' enables visitors to assess the experience of the past as an ongoing concern that is not condemned to the mists of time. The displays are not used as excerpts of a distant history viewed from the security of our current era but as part of our shared environment (Brown and Toadvine 2003). Divisions of culture, nature and time are circumvented in this process as moving through the park is an exercise in forming a broader concept of the social (after Latour 2004). The models that line the paths are not regarded as passive objects that illustrate an unknown past; they are experienced as points of connection in the present. Each reconstruction is recognized as something that 'moves, breathes, lives' and asserts the relationship between the human and the nonhuman (Musée-Parc des Dinosaures 2016). The paths through the site affirm the existence of the 'social' environment (see Bannon 2014).

Forming a sense of place in the dinosaur park

Within the dinosaur parks of Europe, a distinct sense of place enables visitors to regard their situation within the world from an entirely different perspective (after Schütz and Luckmann 1973: 15). This sense of place is highly complex, encompassing multiple time periods and brought forth through the model displays of extinct animals. However, it is significant that this mode of being is not based merely on the human experience but with a far wider understanding of the world that is the crux of a hermeneutical phenomenology: 'the world is always the one that I share with Others' (Heidegger 1962: 155). The phenomenological experience of the contemporary dinosaur park is not an affirmation of the sublime but an engagement with the non-human (after Mann 2006: 163). This new relationship is affirmed within the parks through the signs that encourage visitors to take photographs of themselves with the animatronic displays. For example, DinoPark in Prague and the

Land of the Living Dinosaurs both provide stations along the trails where 'selfies' with the dinosaur models may be taken. In the former park, such displays are termed 'Photosaurus'. These points of engagement place people within times and contexts beyond their own, which destabilizes traditional forms and provides the basis for an alternative society to be considered (see Bannon 2014: 30–31). Humans are no longer the sole interpreters of the environment but part of a wider ecosystem that also possesses the agency to interpret (after Gurwitsch 1977).

This point of engagement can be seen in the Parco dei Dinosauri in Castellana Grotte, Bari, southern Italy. Visitors are guided through the arrangement of dinosaurs and required to contemplate past, present and future: '[The park] will take you back in time, but also forward, to provide you the opportunity to learn about animals that have given life to the fauna that now populate our planet' (Parco dei Dinosauri 2016). Similarly, the Parco della Preistoria in Rivolta d'Adda, Cremona, Italy, evokes the same distinctive sense of place through its arrangement of dinosaur models and dioramas. Over twenty statues have been organized within the site around a parkland trail (Parco della Preistoria 2016). Surrounded by the models of the past, the visitor engages with a world both connected to and different from their own yet in the company of others, both human and non-human (Schütz and Luckmann 1973: 15). Therefore, the engagement with this natural heritage is not conducted in isolation; it is always conducted with regard to the social environment (Latour 2004). The process can be seen in Cardo Land, in Chamoux, Yonne, central France. Within Cardo Land, dinosaur models from various eras are presented with audio recordings of animal noises to immerse the visitor in the sense of an alternative being-in-the-world with others (Cardo Land 2016). Within the Parc Préhistorique de Bretagne in Malansac, visitors are guided along trails where thirty scenes of model dinosaurs are arranged within a 'natural' setting as 'they have never seen it' (Parc Préhistorique 2016). Therefore, no matter the scale or impetus for development for the dinosaur parks – whether scientific discovery, tourist initiative or private ambition – these locales always provide a social engagement with the past; an act of participation that necessitates a consideration of the lives of humans and non-humans alike (Latour 2004). The capacity of sites to create this sense of place is formed through the arrangement of trails and displays that form locales where long-held notions of time, space and homocentrism are transformed and an alternative vision of society can be considered (after Bannon 2014). The models provide points of connection with an alternative environment and a means to regard the world beyond the human; or, a being-in-the-world that I share with others (after Heidegger 1962).

Within these venues, visitors are frequently reminded that they are sharing a world with other animals, which disrupts their place as the dominant species on the planet. This is demonstrated in the Land of the Living Dinosaurs (West Midland Safari Park 2016). The site sets animatronic model dinosaurs

in a landscape of rock formations and planting schemes that mimics the various ages of the Mesozoic Era. From the dominant display of a *Tyrannosaurus rex* that watches visitors from a rocky outcrop to the dramatic reconstruction of an *Argentinosaurus*, the site creates a sense of place that alters notions of being-in-the-world (Schütz and Luckmann 1973: 16). The sense of being-with-other-times and -other-species as a process of being-in-the-present is used as part of this natural heritage throughout Europe's dinosaur parks. It is in such confrontations with humanity's place within the world that dinosaur parks reveal the arbitrary boundaries between 'nature' and 'culture' within contemporary society and embrace a new frontier (after Harrison 2015). As visitors are placed within a time and place where humanity's existence and position on Earth is made into a contingent issue rather than a certainty, the space for an alternative vision of the environment can be created. The dinosaur models and their recreated landscapes constitute a means of placing humans with non-humans to form an alternative experience of the social (see Latour 2004).

Conclusion

It is through the dinosaur parks that an alternative perception of the environment can be formed where humans can consider their place within the world and their relationship with non-humans. Through engaging with such sites, experiencing the fear, excitement, organization and structure of the parks, visitors can reassess the issues of temporality, space and identity that form the bedrock of modern Western society (Giddens 1990). If a different means of thinking about the human and the non–human is to be formed as a means of addressing environmental concerns, then it must be created within spaces that were not constituted by those same established forms of time and space. What is required is a disruption to those values and a recognition that any attempt to confine society to a set of parameters will be framed by those established practices. As such, the relationship between 'nature' and 'culture' within Western society can be reconsidered on the basis of alternative, fluid and renegotiated assessments of time and space to create different ideas of being-in-the-world (after Laclau 1990: 92). This is the format that dinosaur parks provide; pathways through time and space which through a process of excitement, engagement and playfulness, rather than order and categorization, ask visitors to rethink who they are. Significantly, these spaces form a very different type of place where visitors can observe not the distancing of humans and non-humans but their coexistence. It is within the company of others that a clearer assessment of humanity's role as part of the environment can become a matter of concern and society can move beyond the frontier of established notions of 'nature' and 'culture'. Through the application of a phenomenological perspective, the role of a critical natural Heritage Studies can emerge to address how we can construct an alternative vision of the environment as a social world.

Note

1 Styrassicpark, Bad Gleichenberg (Austria); DinoPark, Prague (Czech Republic); DinoPark, Plze (Czech Republic); DinoPark, Vyškov Zoo Park (Czech Republic); DinoPark, Doubrava (Czech Republic); Givskud Zoo, Vejle (Denmark); Préhisto Dino Parc, Lacave en Quercy (France); Cardo Land, Chamoux (France); Musée-Parc des Dinosaures, Mèze (France); Parc Préhistorique de Bretagne, Malansac (France); Parc Dino-Zoo, Charbonnières-les-Sapins (France); Dinosaurier Park, Münchehagen (Germany); Saurier Park, Bautzen (Germany); Dinosaurierpark Teufelsschlucht (Germany); Dinosauria Park, Gournes (Greece); Dinosaur and Adventure Park, Rezi (Hungary); Parco della Preistoria, Rivolta d'Adda (Italy); Il Parco dei Dinosauri, Castellana Grotte (Italy); DinoZatorland, Zator (Poland); JuraPark, Bałtów (Poland); Park Dinozaurów (Poland); DinoPark, Bratislava (Slovakia); DinoPark, Košice (Slovakia); DinoPark, Algar (Spain); Prehistoric Park and Caves, Réclère (Switzerland); World of Dinosaurs, Kalmar (Sweden); Dinosaur Adventure (UK); Dinosaur Park, Tenby (UK); Dinosaur World, Torquay (UK); Land of the Living Dinosaurs (UK).

References

Bakker, R.T. (1975) 'Dinosaur Renaissance', *Scientific American*, 232(4): 58–78.

Bannon, B. (2016) 'Nature, Meaning and Value'. In: B. Bannon (ed.) *Nature and Experience: Phenomenology and the Environment*. Lanham, MD: Rowman & Littlefield, pp. 53–68.

Bannon, B.E. (2014) *From Mastery to Mystery: A Phenomenological Foundation for an Environmental Ethic*. Athens: Ohio University Press.

Brinkman, P. (2010) *The Second Jurassic Dinosaur Rush: Museums and Paleontology in America at the Turn of the Twentieth Century*. Chicago: University of Chicago Press.

Brown, C.S. (2003) 'The Real and the Good: Phenomenology and the Possibility of an Axiological Rationality'. In: C.S. Brown and T. Toadvine (eds.) *EcoPhenomenology: Back to the Earth Itself*. Albany: State University of New York Press, pp. 1–17.

Brown, C.S. and Toadvine, T. (2003) 'EcoPhenomenology: An Introduction'. In: C.S. Brown and T. Toadvine (eds) *EcoPhenomenology: Back to the Earth Itself*. Albany: State University of New York Press, pp. ix–xxi.

Cardo Land (2016) 'Home'. Online. Available at: www.cardoland.com. Accessed 13 June 2016.

Carel, H. and Meacham, D. (eds) (2013) *Phenomenology and Naturalism: Examining the Relationship between Human Experience and Nature*. Cambridge: Cambridge University Press.

Cohen, E. (1996) 'A Phenomenology of Tourist Experiences'. In: Y. Apostopoulos, S. Leivadi and A. Yiannakis (eds) *The Sociology of Tourism: Theoretical and Empirical Investigations*. London and New York: Routledge, pp. 90–111.

Conan Doyle, A. (1912) *The Lost World*. London: Hodder and Stoughton.

Costa, V. (1994) *Dinosaur Safari Guide: Tracking North America's Prehistoric Past*. Stillwater, MN: Voyageur Press.

DinoPark (2016) 'Home'. Online. Available at: http://praha.dinopark.cz. Accessed 12 June 2016.

Dinosaurier Park (2016a) 'Pädagogik'. Online. Available at: www.dinopark.de/Paedagogik/. Accessed 11 August 2016.

Dinosaurier Park (2016b) 'Über uns'. Online. Available at: www.dinopark.de/Dinopark. Accessed 11 August 2016.

DinoZatorland (2016) 'DinoZatorland: Tourism in Poland'. Online. Available at: www.dinozatorland.polturizm.ru/en.html. Accessed 12 August 2016.

Doyle, P. (2008) 'A Vision of "Deep Time": The "Geological Illustrations" of Crystal Palace Park, London', *Geological Society*, 300(1): 197–205.

Duff, C. (2010) 'On the Role of Affect and Practice in the Production of Place', *Environment and Planning D: Society and Space*, 28(5): 881–895.

Eco, U. (1986) *Faith in Fakes*. London: Secker & Warburg.

Gadamer, H.-G. (1975) *Truth and Method*. New York: Continuum.

Giddens, A. (1990) *The Consequences of Modernity*. Cambridge: Polity.

Givskud Zoo (2016) 'Dinosaur Park'. Online. Available at: www.givskudzoo.dk/en/animals/dinosaur-park/. Accessed 12 August 2016.

Greaves, T. (2016) 'Natural Phenomena: The Birth and Growth of Experience'. In: B. Bannon (ed.) *Experience and Nature: Phenomenology and the Environment*. Lanham, MD: Rowman & Littlefield, pp. 31–43.

Gurwitsch, A. (1977) *Die Mitmenschlichen Begegnungen in der Milieuwelt*. Berlin andNew York: Walter de Gruyter.

Harrison, R. (2015) 'Beyond "Natural" and "Cultural" Heritage: Toward an Ontological Politics of Heritage in the Age of Anthropocene', *Heritage and Society*, 8(1): 24–42.

Harvey, D. and Perry, J. (eds) (2015) *The Future of Heritage as Climates Change Loss, Adaptation and Creativity*. London and New York: Routledge.

Head, L. (2016) *Hope and Grief in the Anthropocene: Re-conceptualizing Human–Nature Relations*. London and New York: Routledge.

Heidegger, M. (1962) *Being and Time*. Trans. by J. MacQuarrie and E. Robinson. New York: Harper.

Husserl, E. (1964) *The Idea of Phenomenology*. Trans. by W.P. Alston and G. Nakhnikian. The Hague: Nijhoff.

Ingold, T. (2000) *The Perception of the Environment*. London and New York: Routledge.

Jaffe, M. (2000) *The Gilded Dinosaur: The Fossil War between E.D. Cope and O.C. Marsh and the Rise of American Science*. New York: Crown Publishing Group.

James, S. (2009) *The Presence of Nature: A Study in Phenomenology and Environmental Philosophy*. Houndmills: Palgrave Macmillan.

James, S.P. (2016) 'Phenomenology and the Charge of Anthropocentrism'. In: B. Bannon (ed.) *Nature and Experience: Phenomenology and the Environment*. Lanham, MD: Rowman & Littlefield, pp. 43–52.

JuraPark (2016) 'Home'. Online. Available at: www.juraparkbaltow.pl. Accessed 1 July 2016.

Laclau, E. (1990) *New Reflections on the Revolution of Our Time*. London: Verso.

Latour, B. (1993) *We Have Never Been Modern*. Cambridge, MA: Harvard University Press.

Latour, B. (2004) *Politics of Nature: How to Bring the Sciences into Democracy*. Cambridge, MA: Harvard University Press.

Luckhurst, R. (2012) 'Laboratories for Global Space–Time: Science–Fictionality and the World's Fairs, 1851–1939', *Science Fiction Studies*, 39(3): 385–400.

Lyotard, J.–F. (1984) *The Postmodern Condition: A Report on Knowledge*. Trans. by G. Bennington and B. Massumi. Minneapolis: University of Minnesota Press.

Mann, B. (2006) *Women's Liberation and the Sublime: Feminism, Postmodernism, Environment*. Oxford: Oxford University Press.
Marshall, N.R. (2007) '"A Dim World, Where Monsters Dwell": The Spatial Time of the Sydenham Crystal Palace Dinosaur Park', *Victorian Studies*, 49(2): 286–301.
Masberg, B.A. and Silverman, L.H. (1996) 'Visitor Experiences at Heritage Sites: A Phenomenological Approach', *Journal of Travel Research*, 34(4): 20–28.
Merkl, U. (2015) *Dinomania: The Lost Art of Winsor McCay, the Secret Origins of King Kong, and the Urge to Destroy New York*. Seattle, WA: Fantagraphics Books.
Mitchell, W.J.T. (1998) *The Last Dinosaur Book: The Life and Times of a Cultural Icon*. Chicago: Chicago University Press.
Moody, R.T.J., Buffetaut, E., Naish, D. and Martill, D.M. (eds) (2010) *Dinosaurs and Other Extinct Saurians: A Historical Perspective*. London: Geological Society Special Publications.
Morton, T. (2007) *Ecology without Nature: Rethinking Environmental Aesthetics*. Cambridge, MA: Harvard University Press.
Musée-Parc des Dinosaures (2016) 'Home'. Online. Available at: www.dinosaure.eu. Accessed 12 September 2016.
Noble, B. (2016) *Articulating Dinosaurs: A Political Anthropology*. Toronto: University of Toronto Press.
Owen, R. (1841) *Report on British Fossil Reptiles*. London: Richard and John Taylor.
Owen, R. (1854) *Geology and Inhabitants of the Ancient World*. London: Crystal Palace Library.
Parc Préhistorique (2016) 'Home'. Online. Available at: www.prehistoire.com/. Accessed 30 June 2016.
Parco dei Dinosauri (2016) 'Home'. Online. Available at: www.ilparcodeidinosauri.it. Accessed 16 August 2016.
Parco della Preistoria (2016) 'Home'. Online. Available at: www.parcodellapreistoria.it. Accessed 21 June 2016.
Préhisto Dino Parc (2016) 'Les Dinosaures'. Online. Available at: www.prehistodino.com/les-dinosaures/. Accessed 2 October 2016.
Rudwick, M.J.S. (1995) *Scenes from Deep Time: Early Pictorial Representations of the Prehistoric World*. Chicago: University of Chicago Press.
Samuel, L.R. (2007) *End of the Innocence: The 1964–1965 New York World's Fair*. Syracuse, NY: Syracuse University Press.
Saurier Park (2016) 'Home'. Online. Available at: www.saurierpark.de. Accessed 13 April 2016.
Schütz, A. (1967) *The Phenomenology of the Social World*. Trans. by G. Walsh and F. Lehnert. Evanston, IL: Northwestern University Press.
Schütz, A. and Luckmann, T. (1973) *The Structures of the Life-World*, Volume 1. Evanston, IL: Northwestern University Press.
Secord, J.A. (2004) 'Monsters at the Crystal Palace'. In: S. de Chadarevian and N. Hopwood (eds) *Models: The Third Dimension of Science*. Stanford, CA: Stanford University Press, pp. 138–169.
Sessions, G. (ed.) (1995) *Deep Ecology for the Twenty-first Century*. Boston, MA: Shambhala.
Smith, L. (2006) *Uses of Heritage*. London and New York: Routledge.
Smith, L. and Campbell, G. (2015) 'The Elephant in the Room: Heritage, Affect and Emotion'. In: W. Logan, M. NicCraith and U. Kockel (eds) *A Companion to Heritage Studies*. Oxford: Wiley Blackwell, pp. 443–460.

Society for Promoting Christian Knowledge (1852) *The Industry of Nations, as Exemplified in the Great Exhibition of 1851*. London: Society for Promoting Christian Knowledge.

Solli, B., Burström, M., Domanska, E., Edgeworth, M., González-Ruibal, A., Holtorf, C., Lucas, G., Oestigaard, T., Smith, L. and Witmore, C. (2011) 'Some Reflections on Heritage and Archaeology in the Anthropocene', *Norwegian Archaeological Review*, 44(1): 40–88.

Tolia-Kelly, D.P., Waterton, E. and Watson, S. (eds) (2017) *Heritage, Affect and Emotion: Politics, Practices and Infrastructures*. London and New York: Routledge.

Uriely, N. (2005) 'The Tourist Experience: Conceptual Developments', *Annals of Tourism Research*, 32(1): 199–216.

Wallace, D.R. (2000) *The Bonehunters' Revenge: Dinosaurs and Fate in the Gilded Age*. New York: Mariner Books.

West Midland Safari Park (2016) '*The Land of the Living Dinosaurs*'. Online. Available at: www.wmsp.co.uk/dinosaurs/. Accessed 20 September 2016.

Wilson, R. (2014) 'Playful Heritage: Excavating Ancient Greece in New York City', *International Journal of Heritage Studies*, 21(5): 476–492.

Wilson, R. (2017) *Natural History: Heritage, Place and Politics*. London and New York: Routledge.

Žižek, S. (2014) *Trouble in Paradise: From the End of History to the End of Capitalism*. London: Allen Lane.

Part II

Exploration

5 The Ogre[1]

Doug Scott, CBE

The Karakoram is a harsh, unforgiving land of rock and ice, further north than the Himalaya, beyond the reach of the monsoon, where the soils are thin and growing things scanty. In the middle of this wilderness there is a one-acre patch of lush green grass covered in spring flowers that are buzzing with insect life by midsummer. It is protected from the mountain winds by converging moraines on two sides and by the rocks below the Ogre's Thumb on the other (see Figure 5.1).

On 21 July 1977, the early morning sun shone down on that little triangle of green, bringing the place to life, with butterflies fluttering about. The snow on the rocks above had started to melt, filling the stream that meandered by our old campsite. The heat from the sun was penetrating my soggy, wet sleeping bag; my aching and emaciated body luxuriated in that steamy cocoon of drying feathers. It was so utterly peaceful. There were to be four more days on that oasis with Clive Rowland taking care of Chris Bonington and myself; Chris with two broken ribs and pneumonia, and myself with broken legs down at the ankles. It was now nine days since I had broken my legs just below the summit of the Ogre (7,285 metres/23,900 feet), and four days since Chris had broken his ribs and contracted pneumonia. Mo Anthoine – who had helped Clive get us off the mountain – was the first into Base Camp; or, rather, the first into where Base Camp had been. He was surprised to find all the tents and everything else had gone, apart from a few boxes of food and a note from Nick Escort and Paul [Tut] Braithwaite: 'You are now seven days late. We presume you are all dead. Gone off to form a search party. Nick.' Mo rushed off to try to catch up with Nick and Tut somewhere between Base Camp and Askole, three days away. His main worry was that word would get out to our wives and families that we were missing, presumed dead.

The three of us were on that triangle of green for four more nights; plenty of time for me to go over events on the Ogre. Early on in the expedition, Tut and I had set off to climb the big 3,000-foot south-west rock prow of the Ogre. We reached the start of the hard climbing and went down for more supplies. On the way back up, pulling on the ropes, a rock came rattling down and ricocheted off the sides of the gully, smashing into Tut's thigh and putting

Figure 5.1 At the base of the Ogre
Photograph by Doug Scott

him out of action. Meanwhile, the other four members of our expedition were making their way up the south-west flank of the Ogre, traversing steep mixed rock and ice ribs. Basically, they were working a shift system – two pushing out the ropes and two resting. Mo and Clive eventually reached the top of the ribs just below the western col of the Ogre before dropping back to their camp. Early the next day Chris and Nick set off with the intention of reaching the Ogre's main summit. After several days of hard climbing, with Nick becoming increasingly exhausted and ill with a throat infection, they abandoned the main summit for the west summit. They also realized they had insufficient climbing gear and gas to sustain an effort on that steep, final tower. Meanwhile, Mo and Clive descended to Base Camp.

A few days later, Tut and I joined Mo and Clive in another attempt at the Ogre but with the intention of climbing the west buttress to the west summit, which would give Tut and myself the technical climbing we were after on this expedition. We reached the big plateau and camped. The next morning Chris and Nick arrived, absolutely shattered from what had been an incredible effort at that stage of the trip, when they were both far from acclimatized. Nick decided to go down and await developments at Base Camp before going home, and Tut decided to join him, having finally decided that his injury would hold back the team. Chris was still fired with ambition for the main summit and persuaded us to give him time to recover before rejoining us at a camp we would establish below a 1,000-foot rock pillar. Eventually Chris climbed up to our camp at the foot of the west pillar, although, as he wrote later, he was 'worried about his reception'.

There had been a lot of tension, especially between Mo, Clive and Chris. It had arisen as a result of Chris and Nick suddenly taking off for the summit at a time when Mo and Clive felt they should continue working as a team in order to achieve better acclimatization and move more food and equipment higher up the hill. Now, though, all that was forgotten. In fact, Chris had earned our respect for having gone all the way up to the west summit of the Ogre, all the way back down to Base Camp, and then, after only a few days' rest, back up to 20,000 feet. What incredible drive!

The next day we climbed up the 1,000-foot buttress and eventually reached good snow on the ridge above. Chris and I spent the night in a snow cave, with Mo and Clive on a snow platform in a bivouac tent. The following day we made the second ascent of the west summit and dropped down to the col between it and the main tower of the Ogre. We dug out a big snow cave at around 23,000 feet. The next morning Chris was up early, with porridge cooked as I awoke. We set off first, with Mo and Clive deciding to come along later as Mo was making a film of the climb. I set off in Chris's footprints across deep snow towards the pillar. He was going very slowly and suggested I should lead up the steep, mixed ground to the left of the main tower. The climbing was quite hard at that altitude – a Scottish grade IV/V for about 400 feet. We then traversed down and across to the final rock wall, which barred the way to the summit – now only 400 feet above us.

We climbed up a corner of wonderfully rough, warm granite for 150 feet, protected from the wind and with the midday sun beaming down on us. It was a very enjoyable pitch at about grade 5.7. I continued to lead, on pegs at first but then mostly free climbing – without equipment – before once again using my equipment as direct handhold and foothold aids, relying for safety on nuts and a peg where the crack petered out. At that point, I asked Chris to lower me down, which he did for 60 feet, until all the rope was out. I could now start a pendulum, galloping across the rock from one side to the other until I could reach another crack system far to the right, which I managed to hook with a sky hook. I was almost in the crack when I slipped out and went clattering across the granite. I pulled myself together, checked my bruised body and limbs were still working, and repeated the process. After three more attempts, I managed to jam my hands into the crack without falling out and climbed up to find a placement for a peg. Once again, the rock was brown, weathered, perfectly clean granite. I climbed without gloves to the top of the wall, where I belayed on snow.

Chris came up after me, full of superlatives about the quality of the climb and the exposure (see Figure 5.2). It was the hardest climbing I had ever done at that altitude. A final snow gully led to the summit, but getting into it meant climbing out of a recess and surmounting an overhang, which I did by stepping onto Chris's back. I arrived on the summit just as the sun dipped below the horizon. I belayed Chris and contemplated being up there in the middle of the Karakoram, completely surrounded by glaciers and many other unclimbed, spectacular granite peaks as far as the eye could see.

Figure 5.2 The push for the summit
Photographs by Doug Scott

No sooner had Chris arrived than we were fixing a long length of tape around a summit block of granite for the first abseil. We were now in a rush to get back to the ice cave, our sleeping bags, and Mo and Clive, who had obviously postponed climbing the Ogre until the following day. I fixed the abseil, roped down, and as I went I realized I could save time by clipping into a couple of pegs Chris had left at the top of the wall. There was also a bit of climbing gear hanging around which I hoped to retrieve. To reach the pegs, I had to tension left across the rocks. As I was nearing the pegs, I put my feet a little higher onto what had been a water streak, but in the cold night air the water had turned into a film of ice. Suddenly my feet skated off and I tumbled out of control, turning, spinning, bouncing across the rock. Fortunately, I had enough sense to put my feet up like a pair of buffers, ready for the impact of the fast-approaching rocks on the other side of the couloir. I slammed into the rocks, lost my glasses and ice-axe, and came to a halt dangling on the end of the rope.

I tried to manoeuvre myself onto a tiny ledge, but I knew something was wrong as soon as I pushed with my left leg. I shifted onto my right leg, only to find that it was severely broken: it crackled near the ankle, which left me feeling quite nauseous. I managed to slot in a nut, hammer in a peg and take my weight off the rope. Chris came abseiling down and I told him I had probably broken both my legs. He told me not to panic and insisted that I wasn't going to die. I replied, 'That's good news!' He pulled the ropes down and fixed the next abseil, disappearing into the gloom below.

I had time to take stock of the situation and realized that the rules for winning had just changed. I never doubted that I would get off the mountain – not with people like Chris, Mo and Clive to help me – but I had no idea how it was going to work out in the end. It may sound rather trite, but I experienced a kind of exhilaration at all the uncertainty. In fact, I was impatient to get moving, not because I was starting to freeze but because I wanted to get to grips with this huge problem and try to solve it. The ropes slackened and I slid down them, with my back against the rock, to join Chris on a snow patch in the middle of the tower. 'So,' I thought, 'I can still abseil.' We spent a very cold night as we had left the cave wearing only jumpers and windproof suits – and underpants and socks, of course! The one advantage of a totally unplanned bivouac without any gear is that you simply get up and go the next morning as there are no bags to pack.

The next day, after several more abseils, we reached the snow and traversed over and into the cave with the help of Mo and Clive. The following morning, after eating the last of our supplies, we exited the cave to be met by a full-blown storm that would continue for five days. We went without food throughout those five days. We had another night in the cave but then just had to move. After several abortive attempts, Clive, in swirling snow, put in a huge effort and climbed up to the west summit, leaving loops in the rope so I could pull myself up. Thereafter, it was all downhill. I found I could crawl along in the snow without too much pain, unless I caught the toe of my boot on a rock (see Figure 5.3).

Figure 5.3 Descending on broken legs
Photograph by Sir Chris Bonington

After another night on the slope, Chris smashed his ribs when he abseiled off an unevenly matched double rope onto a single rope and tumbled down for 20 feet, stopping across a jutting rock. From then on, he was in far more pain than me, not least because he developed pneumonia. On the eighth day since my accident, we reached the glacier only to find that Advanced Base Camp had disappeared, so we carried on down towards Base Camp. The next four and a half miles were the most painful for me, as I had to crawl on granite, gravel and the small boulders of the moraine. I arrived in Base Camp at 10.30 p.m. to find that Mo, who had gone on ahead, had found the note and run down the Biafo Glacier in hot pursuit of our friends. He caught up with Nick and Tut before they had a chance to tell our families and friends that we were probably dead. He then continued on to Skardu to try to organize a helicopter.

Meanwhile, down on the triangle of green on the fifth morning at Base Camp, Nick suddenly arrived with a team of Balti porters and enough poplar wood poles to make a stretcher. The Balti were dressed, as always, in their home-spun tunics and cotton pantaloons, sawn-off wellingtons and rubber sandals. Over the next three days I was always in their midst as they carried me along with great care while Chris walked alongside, coughing up yellow phlegm.

Mo's helicopter arrived half an hour after I was deposited in a field outside Askole. It air-lifted me to Skardu, but the engine failed and we crash-landed just short of the helipad, amongst some rocks. It was so badly damaged that

Figure 5.4 On the summit
Photograph by Doug Scott

it had to be replaced. It was seven days before Chris could be evacuated from Askole. Just as he was reaching the end of his tether, the replacement helicopter arrived and took him all the way to the golf course near the British Embassy in Islamabad.

Thus ended our climb up and crawl down the Ogre – the mountain that nearly ate us (see Figure 5.4).

Note

1 Interested readers can find a more fulsome account of the ascent of the Ogre in Doug Scott's recently published book, *The Ogre: Biography of a Mountain and the Dramatic Story of the First Ascent* (Sheffield: Vertebrate Publishing, 2017).

6 Kelakua

Robin Hanbury-Tenison, OBE

I wanted to spend forty days and forty nights alone and so, in February 2003, I went to a place I knew deep in the Sahara to walk, to think and to write.

The camels arrive like an armada, sailing over the nearest dune, past the tree under which I lie waiting; and they hunker down. Fine beasts, fat and with proud humps, they are the best I have ever seen. Kelakua, my old friend and experienced *méhariste*, or 'camel man', gives a great shout of Tuareg greeting as we embrace. Mohammed Ixa, the Tuareg chief who has brought me here, drives away. I will not see or hear another vehicle for forty days.

We are in the remotest place on the dry surface of the earth: the epicentre of the Sahara Desert, which is about the same size as the United States of America. It is one of the hottest, driest and least forgiving environments. It is said the Tuareg, who live here, always save three dates when they are travelling; if all else fails, a man can survive on them for nine days, eating a skin one day, the flesh on the second and sucking the stone on the third. On the tenth day, he will die, but by then he will almost certainly have walked to safety.

No one knows exactly where the Tuareg came from or when they arrived in the central Sahara. They are a Berber people, and consider themselves white, although their skin is burnt dark by the sun. Their babies are born snow white. They have their own language, Tamashek, and their own script, Tifinagh, which is related to the ancient Libyan script and has been used for at least 2,000 years. There are twenty-five consonants represented by lines and dots. The Hoggar Tuareg write from right to left, like Arabic, the Aïr ones vertically, from the bottom up. Their defining attribute is the veil worn by the men but not the women. Traditionally dyed with indigo and made from Sudanese cotton, it gives them one of their popular names: the Blue Men of the Sahara. It has a complex role in their society and its origins remain obscure. Some say its use is hygienic, protecting the eyes from sun and sand and the face from dehydration, but this does not explain why women and children do not wear it; others maintain that its use is symbolic, protecting the wearer from the evil eye and concealing the mouth, which should never be exposed before women and strangers. It adds to their mystery, since one can travel for months with Tuareg companions and never see their faces.

The Tuareg are overtly exotic in their attitudes and physical presence. The period of their conquest by the French and final subjugation to the new, independent regimes as well as their fortitude in surviving an unprecedented series of droughts is a story not widely known. Now, travellers at last have the chance to visit their homeland, some of the most ravishing and spiritually uplifting landscape on earth. The fact that the Tuareg have survived and are cautiously rediscovering their identity gives hope to other nomadic and pastoral societies.

There are probably well over a million Tuareg scattered over many Saharan countries. The largest population is where we are: in Niger, followed by Mali, Algeria and smaller numbers in several other places. Each of the half-dozen main clans has a hereditary king, or *amenokal.* Our clan is the Kel Aïr and I met their *amenokal* in Agadez in 1966. His modest castellated palace in the centre of town had fifteen small, dark windows facing the great mosque, with its characteristic wooden struts all the way up the yellow mud tower. He was a tall, elegant young man in snow-white robes with a fine indigo turban and veil, called a *tagulmust*, behind which I could see soft, friendly eyes as I shook his hand. The Tuareg all kicked off their sandals when he came into the hall and approached him bent double, holding out their clenched right fists before bowing over his extended hand.

Kelakua and I study my maps. I want to go way beyond where small groups of tourists are beginning to be taken to the edge of the Ténéré Desert. Originally, we had planned to go up to the Djado plateau on the border with Libya and perhaps continue to Djanet in the Tassili n'Ajjer, where I made my first camel journey in 1962. That plan has had to be abandoned; recent troubles in the region mean my Tuareg companions would not be able to cross the frontier with me. So, we agree to travel through the northern Aïr Mountains, visiting remote wadis, dried-up river beds. This is a part of the Sahara into which almost no one ever ventures now. Used by the Tuareg during the rebellion in the late 1980s and early 1990s, it is now almost totally abandoned, except for a few nomads with their flocks. Kelakua is pleased. We will be going to places he has not visited for fifty years, but he knows that there will be pasturage for his camels up there and occasional wells.

I'm sitting at my folding table, looking out across the dunes, rocks and thorn trees of the Wadi Tezirzek on the evening before we set off. Achmed, the only one of my companions who speaks French, has buck teeth and takes great care of me. We have travelled together twice before. Now he is cooking supper and talking quietly to some Tuareg men, who have strolled over from their camp. Our things are spread out on the sand and will be loaded on the camels tomorrow.

We will be traversing some of the least known, least visited and remotest countryside on earth. There is a slight danger of kidnap by rebels or murder by 'contrabandiers'.

Around the time of my visit, thirty-two tourists were captured in Algeria by the Salafist Group for Preaching and Combat (Groupe Salafiste pour la

Prédication et le Combat, or GSPC; now known as al-Qaeda in the Islamic Maghreb, or AQIM). Algeria engaged 3,000 civilian and 5,000 military personnel to look for them but it was months before the hostages' cars were discovered where they had been abandoned, booby-trapped and camouflaged under cardboard sheets and rocks; and, although the authorities successfully freed many of the hostages, one German woman died of heatstroke while in captivity. The kidnappers were able to merge into the brisk trade in cigarette-, drug- and gun-smuggling going on in the area, which includes traffic of over 400 four-wheel-drive vehicles crossing the Algerian border each day. Over a decade of civil war in Algeria saw moderate Muslims in conflict with self-proclaimed radical Islamists, and by the end of 2003 over 150,000 civilians, terrorists and security forces had been killed. Algeria's terrorist groups continued to proliferate and on 11 September 2003 Nabil Sahwari, a leader of the GSPC, released a statement declaring that the group operated 'under the direction of Mullah Omar and of the al-Qaeda organization of Usama bin Laden'.

Yet I have been drawn here, not least by the exciting possibility that I will discover some new rock art. Yesterday, as I walked off by myself into the dunes for the first time, I realized it was the silence that had brought me back to the desert. It is tangible. It is why everyone should spend some time in the real desert. It engenders humility. I know I shall be humbled as I enter the deep isolation of the northern Aïr. There, I will be among unimaginably barren mountains; I will feel dwarfed by my surroundings, isolated and entirely alone. I can't wait. As I sit in the darkness under my hanging lamp, while the Tuareg have a most animated gossip around the fire in front of me, the new moon rises to my left and all augurs well.

Socially, the Tuareg are marvellously relaxed yet elegant in their robes of strikingly contrasting colours. They saunter over to talk wherever they happen to find one, whatever one happens to be doing. Five gentlemen, one of whom is Kelakua, just appear, confabulating loudly and heading towards my tree to talk, and they bring some tiny glasses of tea. It comes from China and is called Flecha' – the picture of a bow and arrow is much admired. Very sweet, with a bitter background and aftertaste, it is the most thirst-quenching of all drinks.

The Tuareg have a remarkable way of shaking hands when meeting a stranger, or someone they haven't seen for a long time: a bare touch repeated again and again, maybe a dozen times, with a flourish and much accompanying talk and laughter as recognition dawns. It is a strangely intimate and affectionate gesture between old friends who have not met for a while.

I have always maintained that the most important person to have on an expedition is a local. I have travelled with the Tuareg often before. They are perfect, solicitous hosts. They do not intrude if you wish to be alone to think your own thoughts. They are far more interested in their own affairs than in yours. Once the outline of a plan has been agreed, they know exactly how to implement it. We can never compete with their millennia of adaptation to their particular environment; their practical and instinctive knowledge of the desert. As at home in what appears a featureless wasteland as urban man is in a great city, they are constantly recognizing rocks, landscapes, the lie of the land and so are almost never lost. They would die rather than let anything happen to me.

On my early travels with the Tuareg I was always in a hurry. As I searched frantically for the elusive rock paintings of the Tassili n'Ajjer and the strange rock carvings of the Tibesti Mountains of Chad, and on my first trip to the Aïr Mountains, all in the early 1960s, I felt frustrated at the pace. I was always asking, 'How far? How long?' I found my Tuareg companions' lack of interest in the art of their ancestors and predecessors disappointing. Now that I am older, and I have nothing to prove, I find their rhythm and attitude infinitely reassuring. Each day has its tempo, which is dictated by the needs of the camels, our grotesquely gorgeous companions, upon which we depend for our survival.

Kelakua's camels are gentle, willing creatures. I no longer see their expressions as supercilious, but as concern to do the right thing. They allow the greatest indignities to be performed on them with barely a complaint: such as having the nose held up and all our slops poured down the throat. They seldom wander far at night, in contrast to other camels I have known, and when fetched in the morning they hunker down willingly to be loaded. If something feels uncomfortable, they will mention it by making a hideous open-mouthed groan, but I have yet to see one come close to biting.

Yesterday, Aberog suddenly sat down with me on board. Apparently, something was rubbing his back. Solicitously, Kelakua gathered a soft, sweet-smelling shrub that looked a bit like hare's ears and packed it into the place. They use the same shrub as padding when loading the camels in the morning. The main padding bags are made of soft suede goatskins filled with dried camel droppings. Two short ones are laid lengthwise on top and everything is slung across, with lots more shrub wadding. Whichever Tuareg is lifting a load on always gives a great cheerful shout at the moment of heave.

I have been riding in the Toubou woman's saddle. Kelakua leads, with Aberog attached to his camel and the rest of our party strung out behind. I said I felt like his new bride that he had just bought for many camels and was now taking home. This has become a standing joke on the rare occasions when we meet his nomad relations along the way. Although one of the most inhospitable places on earth, the Tuareg are perfectly at home here and, because of that, I feel perfectly content, too. I would not want to go to a place where no one can feel at home. The rocks are so big and strange in the moonlight, making weird shapes in silhouette.

The four Tuareg, sharing the shade, are crouched round their bowl, eating. The camels have been unloaded and are lying down some way away, while their baggage is carefully strewn near us. Except for Aberog, who hobbled back to us, stood for a while beside my table, then sat down right beside me. Perhaps he is lonely.

Kelakua has just given me a strange little sausage, which he has been making for some time, pounding and grinding ingredients together with his pestle and mortar. He tells me it is made of Tuareg cheese, millet and dates, which are ground whole, stones and all. The taste of the 'sausage' is not too bad, although it does have a lot of sand in it, which grates my teeth. Harder to take is the look of it: the shape and texture of a fresh turd. Due to being rolled repeatedly between Kelakua's palms, it is slightly slimy and smells faintly of sweat. It has been made with such evident care for me and is presented with such solicitude for my welfare that there is no question of not eating it.

We are, by any normal standards, living extremely hard. We are in one of the most barren places on earth, where, when the wind blows, sand moves everywhere with it. We are walking or riding our camels up to twenty miles a day and sleeping at night on a thin mat on the hard ground; and in the same sleeping bag and the same sweaty clothes. Yet I feel wonderful. Always, on an expedition, there are irritations. Yet I have none. I have put myself in the hands of experts who know exactly what they are doing and who are, themselves, patently happy to be doing it.

For example, there is Kelakua's 'dawn chorus'. If you imagine the noisiest and most revolting hawk and spit you have ever heard in an Arab street market and then increase it by a factor of ten, you will have some idea of what my friend does with his nose and throat each morning. From deep inside him I can hear the gobbet of phlegm being resurrected until it is finally ejected with a last alliance of throat, voice box, neck and head, to land on the sand some distance away. Now, this could be maddening, especially if the experience were being shared with others who found it offensive. It was also worrying at first, as I thought there might be something seriously wrong with him. To try to help, I sacrificed three of my precious Strepsyl lozenges (honey and lemon flavour), which I had brought against the almost inevitable cold I usually catch in the desert. He accepted graciously, but nothing changed.

As I lie in my sleeping bag, waiting for the sun to rise and take the first chill off the ground at 6.45 a.m., I look forward to my *méhariste*'s accompaniment as he gets up first and blows the embers of last night's fire alight.

Today, Kelakua suggests that I should go with him to fetch the camels and leave the boys to rest after their long walk in the night. They spent several hours searching far down the wadi when the camels wandered off. I teased them that they were looking for the legendary lone Tuareg girls said to inhabit this region. I feel proud to be trusted to share such an important job, so the two old men set off together and I earn my spurs as a chamelier. It is not so different from driving cattle, just much gentler. Everything Kelakua does with his camels is gentle. He always lets them finish what they are eating. He comes up behind them and softly places his hand on their rump, slowly sliding it across to the tail. They respond to this rather intimate gesture. I lead the young white riding camel with a soft back. We have met no one for two weeks.

I walk faster than the camels. Rounding a corner, I see about a hundred goats spread out along the wadi. Soon, I notice some activity behind a tree and see a very pretty girl in a flowered dress with the goats clustered around her. She starts when she sees me, but walks on past. When I say 'Salaam aleikum' she casts her eyes down and does not reply. Achmed says she is about seventeen. She and her sister live here all year long, near the two wells at the mouth of the wadi. Their father tends the family's camels away in the mountains and their mother looks after the sheep elsewhere.

I stop in a patch of shade to write and wait for the others to catch up. When they come on me, invisible behind a rock, Kelakua is upset for a moment, saying that the camels might have been startled and run away on seeing me, but they hardly turned their heads. I watch him intercept the girl. Shyly, and with her eyes still cast down, she pays no attention and concentrates on her goats as our three lusty boys walk past her. When Kelakua calls out to her, she drops her stick and walks over to him. They do an extra-long 'finger-touching' handshake, maybe seven or eight times, and then have a long chat. We carry on.

In the evening I take a late-night stroll along the sandy wadi in the moonlight. Meeting Kelakua returning from checking his camels, I realize that, robed and turbaned, a tall Tuareg looks like an ancient magician. As we meet, Kelakua holds out his arms and embraces me. With a huge grin he then holds his arms up in the air, a gesture we share, having so little common language, and says, 'Greboun, Temet, T'in Galene, In Azaoua' (All this is Kelakua's country. I know it all and have been everywhere)! He is bursting with pride and happiness. Later, he tells me he has not been to most of the places we are to visit for fifty years, having last travelled there as a boy of fifteen. He never hesitates or makes the slightest mistake with his navigation.

The design of the padding sacks for the baggage camels is very clever. Everything is loaded on top of them, fore and aft of the hump, tied together. Each is a whole soft leather goatskin filled with dried balls of camels' dung. They act like ball bearings to match the contours of the camel's back and so avoid rubbing. The dried dung does not disintegrate to dust, as one would expect, but stays rock hard.

The camels shake their heads up and down to get rid of the flies on their noses. Sometimes, when following in line, one will shove his nose right under the tail of the one in front and have a good rub. This gets rid of the flies for a moment, but camels' bums being what they are, it only attracts more flies at once.

Every time we meet, even if Kelakua has only been to check his camels, he will greet me with 'Salaam aleikum' and a handshake. I reply in kind and add my only Tamashek phrase: 'Surufahit, wérésenakh Tamáschak' (I'm so sorry, but I don't speak Tamashek). This always raises a smile or, as Kelakua is Tuareg, more often a whoop of joy.

For the first week or so, I carried a GPS in one of my pockets and constantly monitored our progress. I soon found that Kelakua's navigation is so accurate that this was a waste of time. In mountainous country, he takes into account the contours of the land and with his local knowledge chooses a far better route than any machine could. On the flat, his orientation is spot on every time, so I put away the GPS, except to record the exact location of any interesting rock art I find. Now in my pocket I carry a slim volume of desert poetry, which I read when I stop and try to learn sections by heart.

Midday. Achmed has just brought me four freshly grilled pieces of goat's kidney on a skewer. I defy any restaurant in London or Paris to beat this for flavour.

A very large white eagle with black edges to its wings has just flown silently over us as we saddle up the camels. Kelakua says that this represents very good luck to the Tuareg. Achmed starts to sing an impromptu song in French: 'Each day Monsieur Robin gets up and says, "On y va."' We don't know where we are going, but Aberog leads the way. I was proud of my camel's name, which I assumed meant he was special, but I learned today that it simply means 'greyish'. The legendary camels of the Aïr Tuareg are white. Mine is off-white, but he is strong and good natured.

I walk for three-quarters of an hour up the rocky Wadi Zeline to the famous *guelta*. This is a proper one. Deep in a rocky fissure, it must be wonderful to swim here when full and fresh. Now it is less than half full and a bit unappetizing, with a good deal of dubious stuff floating in it. But this is the water we will be drinking for the next few days. Rain is expected only every two years or so. Last year there was no water here at Zeline, or at Taghajit or Tamjit, so we could not have made this journey. If there is none for a second year, then things get serious. This happened in 1974 and 1984. Kelakua and many others had to go to Djanet and Tamanrasset to find work, and many camels died.

Tamashek is an odd language. Through muffled ears it sounds like English. I can't understand a word, but the boys chat away animatedly all the time and it seems as though they are discussing world affairs or the latest football results. In fact, I am persuaded that their subjects are really much more mundane matters to do with camp life. Just now, half asleep, I could have sworn I heard one of them say, 'Growing up in the Scottish countryside, of course.' It was just my imagination, but such verbal illusions are not generated by Arabic. Nor are there the accompanying gutturals and aspirants one associates with that language. Those are left to Kelakua, hawking in the background. Nor is there any trace of European languages, like French, Spanish or German. It is just a pleasant, faintly familiar background murmur, which accompanies the music of the spheres.

It is 6.30 a.m. and already everyone is up. The camels have been checked and are grazing nearby and we are eating breakfast. It looks like we are going to cover a lot of ground today. At dawn the glorious sickle moon and the morning star are side by side in a duck-egg sky with an ochre mountainous skyline. This turns through pale tangerine to blue grey as the light strengthens.

Nothing is wasted as we travel. All our dregs – tea-leaves, stew, washing-up water – are gently poured down favoured camels' throats each morning. It is a pleasure to watch Kelakua approach his personal white camel, Abzaw, who clearly loves him, with the tin can of slops. The camel extends his neck with a sweet expression, almost opens his mouth like a bird in the nest, but actually Kelakua does it gently with his left hand, and the sloppy liquid is poured down the animal's throat. There is no objection of any sort. Kelakua's son is doing the same with his camel.

Sometimes unexpected little delicacies are pressed on me. Obscure bits of goat are passed over as they emerge from the cooking over the fire. I have just been given a milky dish of dates, millet and goat's cheese, all stirred into some dried milk and water. This was followed by another of Kelakua's little turds, which I am starting to enjoy, aside from the look of them … and the sand.

I sit and wait for my companions on the top of a small hill. Looking back, they appear over a ridge and approach slowly. As I watch, a sandstorm develops in the desert behind them and moves this way. Gradually, it overtakes them, and they disappear in a cloud of swirling sand. They just plod on, emerging again briefly through the haze and then vanishing again as the storm hits me. I put my head down and wait for it to pass. Then I cut across to rejoin them and to reassure them that I have not managed to get lost. Kelakua navigates accurately and confidently in a die-straight line.

Today has been a fairy story. Like all Tuareg I have known, Kelakua has seemed extraordinarily unselective when we have visited good prehistoric sites, like Iouelene. He makes equally enthusiastic sounds when we find crude stick men and camels as when what I consider great art appears. I was, therefore, not overly hopeful this evening when he said he remembered seeing some good pictures near here as a child. However, as one of my little fantasies about this trip has been that I might find something new and interesting, I have been encouraging him.

The place is the Wadi Sumat, about ten kilometres north of Tesseroukane, our next water site. It is near Mamanet, one of the most famous rock art sites of the Aïr. David Coulson advised me to go there. The Wadi Sumat is a delightful spot with lots of vegetation in a fine setting with good views.

At 4 p.m., Kelakua and I, the two old men, set off together up the wider wadi on the right; the real Wadi Sumat. There were lots of engravings on the rocks on both sides of the valley and, among the more recent graffiti, some nice old animals. I was quite pleased after an hour and a half, so suggested we turn for home. But Kelakua insisted that we should go on a bit further. This was unexpected, as most Tuareg men become bored quite quickly while I search enthusiastically. So, we continued, and around the very next corner we came upon it: a big rock with two mother and baby giraffes, which were just catching the setting sun. There was no question but that these were the real McCoy. They are not as large as the famous giraffes at Dabous, only about a quarter life-size, but I would swear they are of the same era and style. The head of the main mother is somewhat defaced, and she may have been engraved over an earlier antelope or oryx, as there appear to be horns protruding from her head. The ensemble is exquisite and just as Kelakua remembered it. I was thrilled, as much for him as for myself. I told him to recommend them to future tourists, if any were ever to come this way. The animals should be seen, I believe, as we found them: in the light of the setting sun. I was very happy and ready to return to our camp, but Kelakua, whose energy constantly amazes me as he is my age (sixty-seven), insisted we should cross the wide wadi to look at something on the other side. Sure enough, there was another mother and calf giraffe. It was larger than the others and well done, but a little too 'Disney' for my taste. We hurried back in the dark, ecstatic.

This evening I am talking with Kelakua and Achmed about the recent history of their people. The *amenokal*, the king of their clan, whom I met in Agadez when he was a young man, is still their titular ruler and much liked, although he is not a full-blooded Tuareg. They decided long ago that the best way to break the cycle of feuding and fighting, which had crippled their society for generations, was to appoint an *amenokal* from outside. And it seems to have worked. The descendants of the original *amenokal*, including the present one, are the central population of Agadez. They speak a dialect of Hausa but are accepted as Tuareg.

The Tuareg of the Adrar des Iforas, based at Kidal, have never accepted foreign rule. They are now a wild and unruly bunch, by all accounts, living largely from smuggling. Today, the direct route from Kidal to In Salah is much too dangerous. The Algerians use planes to bomb anyone who tries to smuggle goods that way. So, to reach Ouargla and In Salah, the main outlet for smuggled goods, they cross Niger and the Aïr Mountains, enter Algeria by the northern desert and circle back. It is a hell of a long way around and many people have died on the journey. The predominant cargo is cigarettes, the most smuggled item in the world, exceeding heroin and cocaine. The reason appears to be that Philip Morris dumps stock in West Africa, where it is bought at rock-bottom prices by the smugglers. If they reach Algeria or Libya, they can resell the packs for four or five times the cost price. A single successful journey can fund a smuggler's retirement, or perhaps buy new weaponry for a whole rebel division.

Kelakua tells me that the tree above my head has leaves which, if ground with a pestle and mortar, will heal wounds. He says that there are cures in the desert for everything if you know what to use. In my whole life, I have never felt safer, more wrapped in friendship.

Suddenly, after forty days, it is over and Mohammed Ixa arrives in his pickup to take me to the Algerian border and so home. We all embrace, swear eternal friendship, promise reunions. Then we leave. I wave to Abarog as he grazes near the camp and look back to see my friends shrinking to invisibility, waving until we are out of sight.

To go forty days without a hint of a cold or a headache or any of the other minor ills that afflict us daily, but which we hardly notice, is remarkable. To be intensely aware of one's own health and well-being is rare, too. I do get this feeling at home, but not constantly, as I have done here. Part of it is the intimacy with which I have lived with nature. It is very unobtrusive, nature in the desert, at first. There appears to be so little of it: just sand, a few scrubby bushes, occasionally some little birds. That seems to be about it … at first.

Then the power of the silence and the space begins to take hold. The intensity of presence in this random but unique spot in the middle of an infinite expanse. Each little flower really was born to blush unseen, and waste its sweetness on the desert air if I had not happened along. Every fly is here because that is where it is. Each breath of wind blows for a moment through this particular thorn bush in its own special way. Sounds, when they do happen, are swallowed up by the silence, which is ever present: the crunch of a camel grinding up thorns, the clang of a cooking pot as it is packed away. This makes us more aware of everything; perhaps even of ourselves. Perhaps that is why I feel so well.

I have always believed that the more I integrate myself into my environment, the kinder it will be to me. Eventually, nature will return and take over the wasteland we have left behind. Here, it is impossible not to be part of the rhythm of life and to move at that pace because any other course of action would be fatal. There is a little water and some grazing here and there, and the tempo of progress is dictated by the needs of the camels. We could hurry, or we could linger; but the logic of our headway is our rhythm. This, too, has contributed to my feeling of well-being and the sense of rightness of what I am doing. With no contact whatsoever with the outside world, and only very occasional meetings and long talks with the handful of other nomads who spend virtually their whole lives here, and are none the poorer for it, we could be on another planet. This one is hard but endurable. I would love to be able to water it gently. For these nomads, that would be like winning the lottery. Only, instead of changing their lives, it would simply make them even better. I say *even* better advisedly, because living right on the edge, as they do, seems to give those who do it habitually a strength and inner security that we, with our sophisticated society's outer trappings of security, lack.

I don't know if I am any wiser after my forty days in the desert, but I strongly recommend this way of searching for the meaning of life. If nothing else, it lightens the search.

7 'Challenging demoniacal beings'[1]

Extinction, materialities and the mortal frontiers in Alexandra David-Neel's journeys in the Himalayan highlands

Hayley Saul

Introduction

After years of travelling as a Buddhist scholar and pilgrim throughout Asia, Alexandra David-Neel and her adopted son, Yongden, found themselves crossing a remote series of mountain passes in the Tibetan region of Po on their way to Lhasa. The year was 1922; they had all but run out of food, their clothing was melting from their bodies in tattered decay, and deep snow was obscuring their path. The tramping was arduous: Alexandra's feet were a mess of sores where the frozen snow had seeped through her worn soles, and Yongden had sprained his ankle in a fall down a ravine. This was not the first or the last time that death would creep into view. And yet, remarkably, their existence at this mortal frontier only seemed to bring further calm, as Yongden remarked to the female-lama, 'I know you do not fear death, and I do not fear it, either' (David-Neel 1969: 162).

That sentiment has stayed with me since I first read David-Neel's *My Journey to Lhasa*, and in my admiration for the lady-lama's adventures, and her early ethnographies of Tibetan culture and occultism, I sought to emulate her fearlessness in travel. But the significance of this simple statement challenges me in new and acutely emotional ways now. In the 2015 earthquake in Nepal I passed a threshold, a frontier, where I expected to die; it seemed inevitable, I just didn't know how it would happen. The statement now challenges me mostly because memories of that earthquake arrive with *acute* fear. In the aftermath, what trauma has taught me is that retracing former identities, having breached that mutable threshold of mortality, is impossible.

This chapter, then, is a further conversation with the frontier of mortality, through the pilgrimages and experiences of the lady-lama, Alexandra David-Neel. In it, I probe at the threshold of death from a safer distance. From corpse-eating and inviting flesh-eating demons to feast, to zombie reanimation and magical possessed daggers, we discover that Alexandra's concept of the mortal frontier was mediated by encounters with materialities that taunted the extremes of affective experience. Using a blend of Dharmic philosophy and Anthropocene debates, I claim that 'matter comes to matter'

(Dolphijn and van der Tuin 2012) at this frontier of mortality as a way to search for new patterns of thinking–being that emulate those of death. In delivering the five preliminary practices of Tibetan Buddhism, the meditation master Geshe Rabten (2009) identifies amongst them a critical need to contemplate and confront the inevitability of death. The possibility of extinction and the certainty of impermanence are foundational realizations for spiritual development. Truly appreciating what it is to die, navigating those fears, and succumbing to a type of confident acceptance and calm resignation come with practising mortality. By this I mean that Tibetan Buddhism nudges the adherent towards experiences that elicit both a sensorial contemplation of, and an embodied confrontation with, their own transience.

My concern is with a branch of those experiences where death itself is *simulated*, often prompting the devotee into jarring and affectively charged realizations. Such encounters with this morbid frontier are almost always mediated by materialities. Either objects behave as 'tools' to direct the possibility of certain spiritual outcomes, or the corporeal body itself is woven into dark rituals to elicit terror, disgust and vulnerability in practitioners. The late-nineteenth- to early-twentieth-century Himalayan travels of the tenacious female explorer Alexandra David-Neel (henceforth ADN) pulsate with records of these death simulations. Her pilgrimages in India, Tibet and Nepal over the course of fourteen years offer rare glimpses of popular religion and oral histories from the turn-of-the-century Himalayan highlands. One might even argue that her archive is one of the earliest and most comprehensive accounts of non-orthodox, local religious articulations from a time when Tibet was still isolated from foreign *philings.* [2] Using these experiences as an evidentiary basis, I ask: what is the role and potential of materiality in simulating death experiences in the cultural pilgrimages of ADN?

The lady-lama herself seems to have grappled with this question during her travels, which began in earnest in 1911 and proceeded, uninterrupted, for the next fourteen years. By her own admission, she did not venture to Tibet for the purpose of witnessing miracles; rather, she was intent on researching the expression of Buddhism as Lamaism (David-Neel 1965). But her well-travelled researches led her to contextualize these understandings with experiences amidst the Buddhism of India, China, Mongolia, Japan and Korea. In short, she was an accomplished Tibetologist, theologian and early ethnographer of the Himalayas. Of her writings on Tibet, this chapter draws mostly on three volumes, the observations for which were compiled during her travels between 1911 and 1925: *My Journey to Lhasa* (David-Neel 1969), *Magic and Mystery in Tibet* (David-Neel 1965) and *Immortality and Reincarnation* (David-Neel 1978). The morbid mimicry that she recorded in each, as a potential of materialities, is an object of terror for human characters on their journeys but also harnessed by others to explore more deeply the extinction–transformation–creation entanglement that lies at the heart of what, in the West, is otherwise a death/life binary. Because of this philosophical compulsion to be inspired by death, Tibetan Buddhism can rightly claim to communicate a sophisticated understanding of the nature and experience of dying – an antidote to more prevalent dualistic perspectives. Indeed, the venerable Guru Rimpoche, the eighth-century AD master who is credited with introducing Buddhism to Tibet, even taught a comprehensive 'manual' for how to live and what to expect during the disclosing of death, known as the *Great Liberation through Hearing the Bardo* or, more colloquially (through mistranslation), *The Tibetan Book of the Dead.* But how does materiality simulate death? What are the desirable revelations to be found in these practices? And how can these Dharmic forms of thinking–being guide humanity in an Anthropocene epoch that is redolent with extinction?

To answer these musings, I take as my starting point that death in Tibetan Buddhism, both popular and orthodox, is a process of unfolding. Through ADN's accounts of materialities that simulate death experiences, I explore what it is to die: the aspects of death that bleed into 'life'; the permeability of the mortal frontier; and the affective sensations that mediate the life–death metamorphosis. I then invert my gaze to reflect upon what those death-like experiences can teach us about materiality (and materialism) at the frontier of extinction. In doing so, this chapter seeks to resource Anthropocene debates through these transversal conversations with the Dharma, tackling alternative modes of thinking–being about material causality, material agency, the morality of artefacts and corporeality.

Anthropocene and the Dharma: a conversation about extinction

As humanity topples into the new epoch of the Anthropocene, it has become ever more important not just to pose difficult existential questions, but to drive new ways-of-being whilst searching for answers. For instance, the environmental and climate crises and their anthropogenic causes currently operating at global scales (Steffen *et al.* 2011) have stimulated otherwise taboo politics about population control. A few scholars now controversially ask whether the time has come to encourage control of birth rates, especially in countries like the United States where affluent materialism means that, by conservative estimates, a single child has an environmental impact equivalent to at least several dozen Bangladeshi children (Ryerson 2010; Buell 2016). At the other end of the mortal coil, life-expectancy averages around the world have more than doubled over the past two centuries (Oeppen and Vaupel 2002). Similarly, difficult questions are posed by extinction, the rates of which, in vertebrate species, have increased one hundred times over the last century (Ceballos *et al.* 2015). Considering these figures, we are left with a pressing conundrum: we have no more moral authority to extend life than we do to shorten it. Yet, we expend a disproportionate amount of energy as humans on perfecting the former and eradicating the latter. To consider controlling birth rates is virtually a social taboo, but so too, for many, is conceding the inherent ephemerality of a life-principle. Since death began its gradual retreat around 1800 in the wake of epidemiological advances, and conceded an extra thirty years for the average lifespan in the twentieth century, longevity has become an industry bordering on a cult. Coughlin (2017) now estimates that the 'longevity economy' – the unique arrangement of products and services desired or required by an expanding over-fifties market – is worth $8 trillion in the United States alone.

It is not without some dark irony that the human pursuit of extended mortality is a driver of its own extinction threat. The economics of staving off death cannot be divorced from wider materialistic preoccupations as we continue to fashion age itself into a disease that requires potions and elixirs to suppress its wrinkly symptoms. Such is the momentum with which the human species is proceeding towards extinction that it would seem to possess a sublime and yet terrifying gravity. And yet, in the burgeoning populations of the developed-world Anthropocene, the 'materialities of death' are overwhelmingly those that seek to *obscure* its existence. So, it is fitting that we strike up debate between the Anthropocene death–extinction entanglement and Dharmic revelations about the plasticity of our mortality. For, in the Dharma, the materialities of death are manifold, and, more importantly, the role of those materialities in mediating death is far more complex than simply acting to hide it. Instead, as we shall see, form and substance goad actants into diverse deathly experiences: they urge self-grief and they incite disgust at our own corporeality. In doing so, these materialities begin to undo the rift between the death/life binary, but they do so by powerfully harnessing affective experiences of the 'in-between spaces' between death and life, rather than abstractly deconstructing the binary (il)logics.

One of the core recognitions of this chapter, and of the Anthropocene more broadly, is the failure of postmodern aspirations to bridge ontological dualisms that ultimately stem from a Cartesian root. As Dolphijn and van der Tuin (2012: 97) note:

> Prioritizing mind over matter or culture over nature is a transcendentalizing gesture following humanist and dialecticist thought. It posits postmodernism as overcoming the flaws of positivism, and social constructivism as overcoming biological determinism. As such, the gesture is predicated upon sequential negation.

In other words, critical denial of positivism and other modernist perspectives itself reinforces a dualistic before/after, weak/strong, false/correct sequential dialectic. For this reason, there is a concerted focus on non-linear reasoning in Anthropocene scholarly trajectories as a remedy for the failings of progressive logics. This is what Dolphijn and van der Tuin (2011) term 'transversal theory': namely, that reaching for non-dualism means pushing conversations between binaries to extremes, a process that exposes the power/knowledge impetus that generated them in the first place. New knowledge is not a foregone conclusion resulting from an inevitable sequence of thought, or affirmation of the new through negation of the old; rather, knowledge is created transversally as inspired by human, non-human and more-than-human realms through time. The effect is immanent thought (Dolphijn and van der Tuin 2012). The Dharmic practices documented by ADN are a *par de force* for transversally pushing the death/life binary.

Tiptoeing over a threshold and staring down death: traumatized corporeality

First, I confess that such musings about modes of simulating death, and how materiality can play a role in such a confrontation, stem from my own travel experiences in the Himalayas. In 2015 I endured an earthquake in Nepal (Saul and Waterton 2016). Now, I am driven by an immanent, affective need to 'get close to' my mortality, the anticipated outcome of which remains, as yet, amorphous and unclear to me. But these personal reflections are an affective staging post for the enquiries contained in this chapter, and I attempt to use my experiences at the mortal frontier to attune to the revelations instructed by ADN, and more generally in the Dharma.

The first thing I noticed as the tectonic plates began shifting was not the shaking of the ground; it was the cliff face of the opposite river bank peeling away like an orange, slithering into the water below in a bloom of dust and splintered, rocky shrapnel. In that moment, I wasn't scared. I wasn't even scared when the ground started pulsating so violently that our feet were swiped from under us. Landslides of ripped trees and sharpened boulders that were at least thirty metres wide cascaded with seeming liquidity down the mountains to our left and right. The thunderous sound of hundreds of cubic kilometres of rock tearing and clashing was deafening in a way that did not just assault the ears but hummed with a menacing, low-pitched rumble through the whole body. I wasn't scared because I was concentrating. In hindsight, I realize I was in a state of absolute focus. Somehow, I knew to expect the cascade of boulders that started plummeting from above, dislodged from the granite peaks that loomed up to seven kilometres vertically. Sure enough, after a delay that allowed them to pick up speed, massive stones tumbled towards us. Given their enormous size and the speed at which they were travelling, as well as the fact that in some instances they brought with them entire trees that they had snapped on the way down, I fully expected to die. I surveyed this mortal frontier, noting its seeming inevitability. Though in our dualist ontology where mind is 'captured' *in* body one might expect that the point of imminent extinction would be endured primarily as a corporeal concern, during that trauma my body 'disappeared', or at least changed to become unrecognizable to 'me'. Such was my intense focus on my surroundings that the corporeal 'I' momentarily stretched out to become something I can explain only as 'awareness'. So, instigated by the trauma of non-existence, my very corporeality seemed altered to *become* geoforms, ripped trees, violence and movement.

These experiences are reminiscent of the Chöd ritual, which ADN observed as she camped in the grasslands of northern Tibet. Such expansive solitudes were home to few inhabitants, with the exception of a cave-dwelling lama and his *trapa*s (monks of lower rank), and the lonely aspect served the affective summonings of the ritual. During Chöd, the practitioner seeks out a place with an atmosphere of foreboding, in ADN's case a cemetery where a burial had just taken place. Crouching behind some rocks under cover of darkness, she witnessed the *naljorpa* (Tantric practitioner) calling hungry demons to feed on his flesh. In the *naljorpa*'s experience, a feminine, wrathful deity arises before him and severs his head with her sword:

> then, while troops of ghouls crowd round for the feast, the goddess severs his limbs, skins him and rips open his belly. The bowels fall out, the blood flows like a river, and the hideous guests bite here and there, masticate noisily, while the celebrant excites and urges them with liturgic words of unreserved surrender.
>
> (David-Neel 1965: 144)

The shrieks of the *naljorpa* testified that he *felt* himself being eaten. We can look upon his simulated demise as a sacrifice, an act of ultimate compassion and selflessness that placated the spiritual embodiment of anger through a feast of corporeal materiality. He fed her his identity, or at least its physical aspects. Consumption of his own body and the gruesome culinary visualizations that the *naljorpa* viewed play upon disgust reflexes; the sense of compassion aroused is proportional to the significance of the sacrifice offered and disgust heightens the sensation of significance.

What more could be offered up than his own carcass, his extinction? To ask such a question leads on to a more complex realization that if there *is* no more we can sacrifice, then there may be limits to the extent that we can be ethical actants in an era of extinction. How do we live more ethically than we die (or at least *as* ethically)? In her meditations on 'death as recycling', Val Plumwood (2008: 325) challenges us to envisage our death as a salubrious gift, 'a nurturing, material continuity/reunion with ecological others'. Though bodily gifts are, indeed, demonstrations of compassion that go some way to redressing the 'alienation from the earth' that has allowed us to pursue its ruination, are such material gifts enough? In other words, are we limited as ethical actants by our corporeality? Is the *most* we can sacrifice compassionately our own bodily materiality? The Chöd ritual would not have us limit ourselves there. The 'elation springing from the idea of sacrifice' (David-Neel 1965: 144) is extinguished as the now *naljorpa*-corpse continues the rite, visualizing himself oozing from black mud as an irrelevant heap of charred bones. His realization is that his sacrifice is an illusion; he has nothing to gift because he *is* nothing. His corporeal form has already transformed, through death, into something that is not 'him', and can therefore not be attached any more to the intention of 'gift'. The ritual use of a *kangling* (a flute-like musical instrument made from a human thigh bone) in Chöd to summon the hungry demons is suddenly brought into new relief: a reminder that we all end up as something transformed, our corporeality less significant than we might otherwise believe. Ultimately, our material purpose is infinitely reappropriated into the cyclical life–death–recreation entanglement.

If, as this suggests, the sacrifice of death is not our most (or, at least, only) compassionate gift, how do we then supplement this by *living* as ethical beings? Does the material body define the limit of our ethical possibilities? The Dharma would answer 'no'. The *naljorpa*'s vision of his bones mired in sludge is an affective prompt that even if we materially transform beyond recognition of our*selves*, even becoming nothing, we are *substantial* in a way that is more than our corporeality. Substance is not reliant on material form, as long as it is causal. This is a root in the Dharmic notion of 'dependent origination': 'Nothing exists in its own right independent of other factors. Things and events come into being only in dependence on the aggregation of multiple causes and conditions' (His Holiness the Dalai Lama 2005: xvi). Substance, therefore, is indivisible from causation. The key realization for our Anthropocene existence is that we are not only ethical beings in a world where cause and effect exist according to linear mechanical possibilities; we are ethical beings in a world where causation is non-linear, driven by attractions, aversions and karmic propensities that might never manifest corporeally. To act ethically in the Anthropocene means seeking to reorient ourselves amidst a world of tangible *and* intangible substance that is created, dissolved, acts and reacts in non-linear and evanescent ways. Out of sight, and even out of mind, is not necessarily out of action. Whilst due consideration is being prompted for us to open ourselves up to more-than-human phenomenal worlds in shaping our eco-ethics (Cairns 2002), what is still missing is an extension into the complex causalities at the frontier with the *extra*-phenomenal world where what 'lives' cannot conform so easily to the rules of the death/life binary.

The lively ethics of animating things

ADN encountered this frontier in realms beyond ordinary phenomenal experience on several occasions. In fact, she went out of her way to gather stories and learn about it. Often the tales invoked the materiality of the body as a medium for encountering the altered capacities of things beyond the mortal frontier. In one such tale, an abbot of Miniagpar Lhakhang near Tachienlu, Chogs Tsang, visited a nearby river late at night to collect a 'meal' delivered from an acquaintance in India. Floating against the current, bathed in a light source that it seemed to carry with it, was a corpse. Chogs Tsang cut flesh from the body and ate it. The flesh, according to ADN, had accumulated 'special qualities' that were acquired through the attainment of spiritual perfection, and these knowledges, ecstasies and supernormal powers could be transferred by consumption. In a real sense, the corpse remains animated by these potentialities, which retain causative power.

The reanimation of the corpse is seen repeatedly in ADN's travels, sometimes in a more literal sense. The lady-lama was once told about the *rolang* (meaning 'corpse who stands up') rite by one who had practised it. The *ngagspa* (magical practitioner) in question lay on top of the corpse in a darkened room, each affixed to the other by the lips. After some time of incanting a magical mantra, the corpse began to twitch and writhe, gradually increasing with intensity until the unnatural symbiosis was propelled around the room, from wall to wall. According to ADN, it was essential that the *ngagspa* should not lose contact with the corpse's lips, for to do so would mean his instant loss of control and death at the mercy of the *rolang*. As the twisting and jerking of the corpse reached powerful levels, we are told, it stuck its tongue into the mouth of the *ngagspa*, at which point he clenched his teeth and bit off the tongue, effectively capturing the potent animative power of the *rolang*. It then fell to the floor and traversed back beyond the mortal threshold.

What binds these narratives, whether they are to be taken literally or metaphorically, is a message that, just as death is transformative, so the corpse contains within it transformative power. This is certainly expressed in these rituals: harnessing the transformative power of death means confronting it and knowing it, to eradicate fear of it. Fear drives actions that mask and distort the imminence of death; our perceived temporality, our false longevity, surely distorts our appreciation for the frailty of other species and our manner of ethical engagement with them. Acquainting ourselves with our own death, subduing fear, is perhaps a methodology for transcending barriers to ethical action. It is telling, I think, that these experiences at the mortal frontier deliberately elicit or play on disgust reflexes and traumatizing sensations. They would seem to point out that traumatic experiences (re)compose our own corporeality by reorienting the body in new patterns of similarity and difference to death. We are led to know death by exploring its various aspects and allowing our lively animation to become enfolded with qualities brought back from beyond the mortal frontier. Empathy with death transforms the ethics of our lively interactions.

We begin to tread on dangerous ground here. ADN was acutely aware of the dangers of these knowledges: to probe beyond the frontier of 'what lives' is to become entangled in the complex, non-linear causation that composes much of the *extra*-phenomenal realm of intangible substance that inhabited early-twentieth-century Tibet. Anna Tsing touches on this idea when she notes that, 'theories of heterogeneity are still in their infancy. To appreciate the patchy unpredictability associated with our current condition, we need to reopen our imaginations' (Tsing 2015: 4). For ADN, heterogeneity poses dangers because it confounds predictability. In seeking to 'know' the mortal frontier, and the unruly processes by which things (ideas, objects, forms) become animated, we expose ourselves to turbulent possibilities for how those lively things might act.

Even solid rock can learn to conspire, as I was taught by the earthquake. Somehow, as the ground shifted violently I found a few brief moments to ponder what it would feel like to be struck by a rock and not wake up again. I remember hoping that I wouldn't be pinned down by a large boulder, unable to move, suffocating under its weight. Instead, I would just go, I hoped. So I tiptoed across that mortal threshold; and even though I wasn't hit by a rock, I was strangely crushed on that day by the very certainty of it. In that sense, the rock by which I would meet death materialized, and it did so with a powerful causality without ever taking on a geological being.

Since the earthquake, a residue of trauma, dragged back across death's frontier, has glazed over my formerly mundane encounters with materiality. To most, my responses would be barely recognizable as something we might neatly term 'post-traumatic stress disorder'. I would like to think that's because I'm in control, but the reality is that matter wields a new type of subtle and insidious power over me. Now, the approaching rumble of a London tube train always makes me claustrophobically aware of how much rock is poised above me. How does it not collapse into that ready-made cavity? I find myself assessing the roof space for resilient hiding spots, prompted by the bass note even before I register the train's approach. The margins of geology's solidity can't be trusted any more. Useless hours have been spent trying to comprehend the mountain's targeting strategy: by what virtue am I still alive, and to what must I commit in order to avoid a violent death in the future? Surely such volumes of matter could be moved only by a wrathful deity or a roaring monster, demanding my supplication? I am jarred awake by the lumbering motion of turbulence on an aeroplane now: the fuselage dips and bobs mockingly, and my body responds with rhythmic pulses of adrenaline as I expect the floor simply to drop away. Worse still, I feel 'phantom' earthquakes; sometimes, at night, they even wake me up. In those moments, I have to stop and send deliberate proprioceptive 'feelers' into the world; and if those feelers don't countermand belief, I have to check online geological surveys for real-time quake data. Should such feelings about the powerful, sometimes dangerous, 'aliveness' of matter be dismissed as anomalous and abnormal?

Precedents for credulity exist in ADN's experiences. Whilst in northern Tibet, the lady-lama encountered a caravan of lamas who were tremulously transporting a *phurba* (ritual dagger). The blade, it was believed, had begun to wreak misfortune upon its owner's monastery, following his recent death. ADN records, 'Two of the three monks who had touched it had died, and the other one broke his leg by falling from a horse' (David-Neel 1965: 133). Strange noises could be heard coming from the box in which it was imprisoned. Fancying her chances of taming whatever causal propensities inhabited the object, ADN sat out the night with the *phurba* her only companion, plunged into the ground beside her. Several hours into the darkness she witnessed a withered figure stooping over the dagger, as if to retrieve it. Having stumbled to her feet to prevent the theft, ADN looked around only to find the plains empty, desolate of occupants. And yet, even as she stood, confounded by the simultaneous presence and absence of the animate being, the dagger was felt, 'feebly moving' in her hand (David-Neel 1965: 134).

What distinguishes the dagger in this tale from other well-documented ritual objects inhering with power (Gentry 2016) is that the *phurba* had become intentioned, almost calculating. This is something of an extension of the moral equivalence of humans and objects proposed by Latour (1993, 2002), who argues that many objects embody and act out moral norms; but even though they are not neutral, they still operate within a human moral framework. That is, materialities do not necessarily have moral agency in themselves; their morality relies instead on being networked amidst humans (Verbeek 2014). The *phurba* that struggled to be free of ADN's grasp, however, would seem to urge us to consider materiality's ability to extend a little bit further, sneaking across the frontier from lifeless to lively. Interestingly, ADN's fable-like explication of Dharmic understandings of *how* objects become animated accords closely with Latour's (1993, 2002) insistence upon a mutually reliant human–object network.

ADN describes three related lively object manifestations: *yiddam*s, *tulku*s and *tulpa*s. All are generated by way of esoteric methods that hone the concentration of an adherent to a point of such focus that it becomes densely matter-like, and variously they may employ a diagrammatic space that demarcates the frontier between animate and inanimate, called a *kyilkhor* (David-Neel 1965). A *yiddam* is a 'tutelary deity', which might appear as a wrathful demon (noting that simple 'good' and 'bad' gives way to multi-causality in the Dharma, so even demons embody important pedagogical messages) and is evoked into existence by a student of some esoteric branches of Buddhism in order to learn from it. A *tulku*, however, falls into a more general category described by ADN as 'phantoms' and might variously appear (and not necessarily purposefully) as landscapes, people, animals and objects. Finally, ADN had direct experience of a *tulpa* – a human-like apparition with animate qualities – as she created one herself.

ADN is cautious about describing in detail the practice of generating a *tulpa* because,

> Endowed with enough vitality to be capable of playing the part of a real being, it tends to free itself from its maker's control. This, say Tibetan occultists, happens nearly mechanically, just as the child, when his body is completed and able to live apart, leaves its mother's womb.
>
> (David-Neel 1965: 280)

This suggests that the lively animation of the visualization is achieved not just because the *tulpa* seizes its own intentionality and phenomenal reality, but because the *process* of traversing the mortal frontier to animation generates sufficient momentum to overwhelm the vulnerable limits of human intent. First and foremost, the message is that actions that (re)form the plasticity of the mortal frontier need to be undertaken with strong intent, meaning being mindful of the complex causalities that may result. Further, though, in the sense that the *tulpa* arises from its interdependence with networks of human morality and intention, the practice strongly resembles the Latourian notion of the meaning and purpose of things: object morality (Latour 1993, 2002). The human invests the *tulpa* with a life-principle, a kind of moralistic energy that is so concentrated that it creates form out of thought.

The divergence occurs, however, at the point where the *tulpa* snatches *independence* from those human networks. Having saddled them for the purposes of corporeal existence, it breaks loose and its morality (intimately linked to its life-principle) takes on a momentum of its own. Often, ADN records, that morality may lead the *tulpa* to injure or kill. Indeed, her own experience of creating one nearly cost her dear. Having shut herself into secluded concentration for several months, 'a phantom monk was formed … His form grew gradually *fixed* and life-like looking. He became a kind of guest living in my apartment. I then broke my seclusion and started for a tour' (David-Neel 1965: 281; original emphasis). The monk travelled with ADN's party, sometimes appearing even when she was not thinking about him. He performed actions of his own volition, admiring scenery and walking beside the group. ADN even *felt* his robes brush past her on several occasions, and others were able to interact with him. But then,

> The features which I had imagined, when building my phantom, gradually underwent a change. The fat, chubby-cheeked fellow grew leaner, his face assumed a vaguely mocking, sly, malignant look. He became more troublesome and bold. In brief, he escaped my control.
>
> (David-Neel 1965: 281)

It was only after several months of concentrated 'dismantling' of the *tulpa* that ADN finally caused it to cross the mortal frontier once more and disappear. But it was 'tenacious of life' (David-Neel 1965: 282) by that point, and did not concede its animate reality without resistance.

What message might we take from this fable, whether read literally or as Dharmic metaphor? ADN would have us be mindful of the plasticity of the mortal frontier, and the dangers of entangling ourselves with that elasticity. In a literal sense, an animate–inanimate dichotomy cannot be assumed, according to the experiences of the lady-lama. In a reality where living and dead are fluid, extinction becomes a less discrete entity. It is a warning to the Anthropocene that extinction 'events' occur with varying degrees of physical tangibility, in varying intensities of form or corporeality. Conversely, we are responsible for amorphous 'demons' that condense into forms, with causal power, without ever necessarily organizing our human intent. For instance, a growing body of research into social media shows a strong correlation between internet use and social anxiety (Caplan 2007), as well as an elevated risk of bullying and sexual predation (Pujazon-Zazik and Park 2010). Though intended as a positive mode of reducing isolation for young people with interests outside mainstream culture (Moreno *et al.* 2009), the unintended contrary effects include increased risk of victimization for females and gay adolescents (Wolak *et al.* 2008). The lively momentum with which the internet has taken form continues to elicit multi-linear effects that transcend the networks of human morality and intention in which the technology first emerged. In many ways, this causal independence might be viewed as stretching the malleability of the mortal frontier, taking on life.

Concluding matter: mortality's mimic

In this chapter I have sought to use the experiences of the lady-lama Alexandra David-Neel as she encountered the various and multiple aspects of the mortal frontier to reflect upon the nature of the death–life–transformation entanglement. ADN encountered the mortal frontier in a range of ways: she witnessed hermit monks sacrificing their corporeal existence to demons, and emerging transformed; she met practitioners who had 'awoken' the dead in order to steal body-parts that dragged 'special qualities' back from beyond the deathly boundary space; she succeeded in creating substance, causal propensity and lively animation within a dense arrangement of her concentration. Out of these enterprises, we come to an awareness of death as an unfolding process, possessing qualities and degrees that are unbounded. The Dharma guides ways to renounce the simplistic notion of a death/life binary and embrace heterogeneity. However, in renouncing such a duality, further questions arise that become important in efforts to navigate the Anthropocene – an era of extinction. How do we live and act ethically in a resource-strained world where one of the most compassionate sacrifices we could give is – arguably and controversially – our own death? The answer is that coming to know the viscerality of death, to feel it and enact (self-)grief, is a way of transcending fear. Fear of death – that is, fear of compassionate sacrifice – is a barrier to ethical action because fear guides us to appropriate the liveliness of earth-others so that we can create the illusion of control. Conversely, though, the Dharma also cautions fear of life. Animation can steal into existence just as lively matter can dismantle and traverse into a silent and unacknowledged extinction.

To act ethically in the Anthropocene, we must consider not just the matter that slithers across the frontier into death/extinction but also the ethics of what we give vitality to. Dangers inhabit these entanglements that make up the mortal frontier because of the multi-directional nature of the death/life relationship. To navigate these mortality ethics, the Dharma warns us to heed dependent origination (complex, multi-linear causality) carefully in our encounters with the death–life–creation frontier. Much as Cameron (2015: 100) concedes that we face 'a future in which the certainties of the past have gone and the future lies before us unknown', the concept of dependent origination maintains that in a reality of multi-linear and complex causality, those certainties of the past never existed in the first place. Calls for an attitude of curiosity or 'open experimentation' (Cameron 2015) in Anthropocene research, however, perhaps do not sit well with an attitude that is vigilant of causal minutiae. Yet, I would argue that the two are not in conflict. Instead, they are two points in a process of acting ethically. Whilst part of overcoming fear means we *must* strike with openness and make 'a stand for certain worlds and for certain ways of living on the planet' (Cameron 2015: 100), what follows, amidst a world that transforms and adapts in symphony with us, must be a dynamic and reactive observance of the lively morality of those decisions. In other words, to live and die curiously also requires that we take responsibility for the consequences of these new, expanded ecologies of beings that occupy the living–dying–transforming flux. And, if the experiences of ADN are any guide, the *extra*-phenomenal world where mind and matter intercept each other is a fertile zone to confront receptively an expanded notion of 'what lives' in the Anthropocene.

Notes

1 David-Neel (1965: 136).
2 During Tibet's isolationist period in the nineteenth century, foreigners were termed *philings*, and would be escorted to the border if discovered upon Tibetan soil. Before ADN's escapades to reach the forbidden city of Lhasa in 1923, a number of Westerners, including Annie Royle Taylor in 1892, Mrs St George Littledale in 1895 and Susie Rijnhart in 1898, had made unssucessful attempts to penetrate the Tibetan frontier (David-Neel 1969). ADN's success at reaching Lhasa is undoubtedly attributable to the fact that her approach was culturally immersive, pioneering ethnographic values of cross-cultural participation as a means to understand an alternative perspective, rather than attempting to impose a Christian perspective, which had been the impetus for the earlier efforts.

References

Braidotti, R. (2010) 'The Politics of "Life Itself" and New Ways of Dying'. In: D. Coole and S. Frost (eds) *New Materialisms: Ontology, Agency and Politics.* Durham, NC: Duke University Press, pp. 201–220.

Buell, L. (2016) 'Anthropocene Panic: Contemporary Ecocriticism and the Issue of Human Numbers', *Frame*, 29(2). Online. Available at: www.tijdschriftframe.nl/29–2–perspectives–on–the–anthropocene/1449/. Accessed 15 March 2018.

Cairns, J. Jr (2002) 'A Declaration of Eco-ethics', *Ethics in Science and Environmental Politics.* Online. Available at: www.int-res.com/articles/esep/2002/E21.pdf. Accessed 12 December 2017.

Cameron, J. (2015) 'On Experimentation'. In K. Gibson, D. Bird Rose and R. Fincher (eds) *Manifesto for Living in the Anthropocene.* New York: Punctum Books, pp. 99–102.

Caplan, S.E. (2007) 'Relations among Loneliness, Social Anxiety and Problematic Internet Use', *Cyberpsychology and Behaviour*, 10(2): 234–242.

Ceballos, G., Ehrlich, P.R., Barnosky, A.D., García, A., Pringle, R.M. and Palmer, T.M. (2015) 'Accelerated Modern Human-induced Species Losses: Entering the Sixth Mass Extinction', *Science Advances*, 1(5): 1–5.

Coole, D.H. and Frost, S. (2010) 'Introducing New Materialisms'. In: S. Frost and D. H. Coole (eds) *New Materialism: Ontology, Agency and Politics.* Durham, NC: Duke University Press, pp. 1–43.

Coughlin, J.F. (2017) *The Longevity Economy: Unlocking the World's Fastest-growing, Most Misunderstood Market.* New York: Public Affairs.

David-Neel, A. (1965) *Magic and Mystery in Tibet.* London: Corgi Books.

David-Neel, A. (1969) *My Journey to Lhasa*, 2nd edition. London: Virago Press.

David-Neel, A. (1978) *Immortality and Reincarnation: Wisdom from the Forbidden Journey.* Rochester, VT: Inner Traditions.

Dolphijn, R. and van der Tuin, I. (2011) 'Pushing Dualisms to an Extreme: On the Philosophical Impetus of a New Materialism', *Continental Philosophical Review*, 44: 383–400.

Dolphijn, R. and van der Tuin, I. (2012) *New Materialism: Interviews and Cartographies.* Ann Arbor, MI: Open Humanities Press.

Gentry, J.D. (2016) *Power Objects in Tibetan Buddhism: The Life, Writings, and Legacy of Sokdokpa Lodrö Gyeltsen.* Leiden: Brill.

Henare, A., Holbraad, M. and Wastell, S. (2007) 'Introduction: Thinking through Things'. In: A. Henare, M. Holbraad and S. Wastell (eds) *Thinking through Things: Theorizing Artefacts Ethnographically.* London: Routledge, pp. 1–31.

His Holiness the Dalai Lama (2005) 'Introductory Commentary by His Holiness the XIVth Dalai Lama'. In: G. Coleman (ed.) *The Tibetan Book of the Dead.* London: Penguin Books.

Latour, B. (1993) *We Have Never Been Modern.* Cambridge, MA: Harvard University Press.

Latour, B. (2002) 'Morality and Technology: The Ends of the Means', *Theory, Culture and Society*, 19(5–6): 247–260.

Moreno, M.A., Parks, M.R., Zimmerman, F.J., Brito, T.E. and Christakis, D.A. (2009) 'Display of Health Risk Behaviors on MySpace by Adolescents: Prevalence and Associations', *Archives of Pediatric and Adolescent Medicine*, 163: 27–34.

Oeppen, J. and Vaupel, J.W. (2002) 'Broken Limits to Life Expectancy', *Science*, 296 (5570): 1029–1031.

Plumwood, V. (2008) 'Tasteless: Towards a Food-based Approach to Death', *Environmental Values*, 17(3): 323–330.

Pujazon-Zazik, M. and Park, J. (2010) 'To Tweet, or Not to Tweet: Gender Differences and Potential Positive and Negative Health Outcomes of Adolescents' Social Internet Use', *American Journal of Men's Health*, 4(1): 77–85.

Rabten, G. (2009) *The Preliminary Practices of Tibetan Buddhism.* Dharamsala: Library of Tibetan Works and Archives.

Ryerson, W.N. (2010) 'Population: The Multiplier of Everything Else'. In: R. Heinberg and D. Lerch (eds) *The Post-carbon Reader: Managing the 21st Century's Sustainability Crises.* Healsburg: Watershed, pp. 155–174.

Saul, H. and Waterton, E. (2016) 'Leaving Langtang'. In: R. Smith (ed.) *The Day the Earth Shook.* St Ives: Story Terrace, pp. 155–166.

Steffen, W., Persson, Å., Deutsch, L., Zalasiewicz, J., Williams, M., Richardson, K., Crumley, C., Crutzen, P., Folke, C., Gordon, L., Molina, M., Ramanathan, V., Rockström, J., Scheffer, M., Schellnhuber, H.J. and Svedin, U. (2011) 'The

Anthropocene: From Global Change to Planetary Stewardship', *AMBIO: A Journal of the Human Environment*, 40(7): 739–761.

Tsing, A. (2015) *The Mushroom at the End of the World: On the Possibility of Life in Capitalist Ruins.* Princeton, NJ, and Oxford: Princeton University Press.

Verbeek, P.-P. (2014) 'Some Misunderstandings about the Moral Significance of Technology'. In: P. Kroes and P.-P. Verbeek (eds) *The Moral Status of Technical Artefacts.* Dordrecht: Springer, pp. 75–88.

Wolak, J., Finkelhor, D., Mitchell, K. and Ybarra, M. (2008) 'Online "Predators" and Their Victims', *American Psychologist*, 63: 111–128.

8 Everest and the Himalaya

The evolution of mountain travel and exploration

Tim Macartney-Snape

Introduction

The industrial age reached its peak towards the end of the nineteenth century, just as the largest blanks on the earth's surface were being filled in. The golden age of physical exploration was coming to an end. The mass migration of people from rural areas to towns and cities meant occupations became decreasingly physical and increasingly sedentary, and more people than ever before were afforded leisure time. Conditions were ripe for the birth of mountaineering as a pastime. Almost simultaneously, on the cliffs of the Lake District in England, Saxony in Germany and the Dolomites in Italy, rock climbing became a weekend pastime for people from wide-ranging socio-economic backgrounds looking to escape into the hills. Taking longer excursions into the grander-scaled Alps was an extension of this, though initially it was wealthier English climbers who went there and employed locals with a natural affinity for climbing, or those who had developed mountain expertise from shepherding and crystal hunting, as guides. By the start of the twentieth century, climbers were fast discovering the mountaineering intricacies of the Alps, but exploration of the greater ranges in Asia and the Americas was only just emerging. Within those rugged landscapes were some of the last tracts on earth that had not had human eyes cast over them. More than a century later, there is still some steep terrain that remains untrodden.

The motivations to climb are many, but underlying them all is an urge to explore not only geographical but psychological boundaries. In my own experience, the exploratory motive has always been a powerful driving force in pursuing my passion for mountaineering and in choosing where to travel in general. Encountering challenging conditions enriches the experience because it opens up the psychological realm of exploration. Adversity cuts through the fog of day-to-day preoccupation and insecurity, bringing focus to the present. The more intense the adversity, the more intense is that focus. It has the power to restore in an individual a greater sense of resilience, strength and confidence and can even rekindle childlike enthusiasm. In this chapter, I use two of my formative expeditions to the Himalaya to provide some insight into exploration and its motivations before touching upon the challenges we face in managing the shrinking arena for remote exploration.

Figure 8.1 Annapurna II
Photograph by Tim Macartney-Snape

10 October 1983

I have one foot placed securely, relatively speaking. Two steel crampon points protruding forward from under the toe of my right boot are sunk into a vertical ribbon of ice. The ice dribbles down a corner of rock that's a major feature of a 400-metre headwall of sedimentary rock, the final barrier to the summit of this Himalayan giant called Annapurna II (see Figure 8.1). Below my feet, two ropes hang loosely, threaded through a few questionably secure anchors to my climbing partners, huddled and undoubtedly shivering on a ledge thirty-five metres below.

My other foot's perch – in the same runnel but higher up – is less secure because the ice shattered on impact. As it is the higher foot, I have to put my weight on it. In my left hand I hold an ice tool, the tip of which rests precariously on a nubbin of rock. Pulling down on it gingerly, I step up, trying to keep my foot completely still as any movement will pry it from its tenuous hold. Another couple of scary moves and I have both feet on a good ledge. Thank God I can rest, my chest heaves and heaves, gasping for breath in the arid, cold air. I place a couple of anchors in the rock, tie myself in, pull up the remaining slack in the ropes, secure them, and call up the others. It's a slow and tedious process.

Arching back and looking up, we see only overhangs. I scan for possibilities. Surely there's a way around – an unseen system of ledges or cracks that will allow us to bypass these intimidating roofs? But on steep ground such as this, nothing is revealed until you're upon it (or sometimes above it). You must fight fear and the barren air that forces you to breathe at four to six times the normal rate, and every move must be calculated as you scan ahead, trying to find a path of least resistance. The summit is the ultimate goal, but you know that what is immediately needed is to concentrate on the next few moves. The inspiration to continue comes not only from a vision of the summit shining in the sun but also from the view that unfolds as I climb. Looking down between my legs, I see the ramparts below, sections of which were once a fearsome mystery and the cause of so much struggle, but now known, familiar, almost like a friend.

I see the steep, snowy ridge where, earlier in the climb, I had set off an avalanche, and relive the terror as the whole snow-slope underneath me had suddenly slid, accelerating towards a cliff edge as I ran on the spot to get off the unwanted ride. Then, further below, the jungle-filled gorge splits the high foothills, snaking its way down to the lower hills. Weeks before, working alongside the local porters who helped to carry our supplies and equipment to the jungle base camp, we had to recut their path up the valley as it was overgrown with monsoon verdure. Their stout, tough stature made it obvious why soldiers recruited from this tribe were held in such high regard by their Gurkha regiments. Together we forded flooded streams, bush bashed, slashed our way through stands of giant stinging nettles, mud and leeches everywhere, and all along they shared some of their jungle lore. They took us to a clearing where the gorge widened to create a little oasis of level land. It was a summer settlement they called Hoga, accessed circuitously during the wet season from a high ridge bordering the gorge, where potatoes, maize and beans grew with fantastic vigour in the rich soil. This fecundity would provide a great boon to our diet over the next few weeks.

Now, that waterlogged air, rich with the smell of decaying leaves and buzzing with insects, is a distant memory. Once we reached the foot of the climb, our porters returned to their village. They seemed to appreciate the challenge we had set ourselves but venturing any further themselves was beyond their world of possibilities. Their daily lives were fully occupied with the myriad challenges of living off the land on the often-hostile edge of human habitation.

The monsoonal deluges have gone now. The crisp autumnal air is clear to the blue horizon. It feels good to be alive – truly alive, all senses heightened, despite the thin air and the nagging fear of being out on a long, high limb, far from home. Consequently, there is always an eye questioning the horizon. Will the weather hold? Just as urgently, will any ledge be wide enough for us to spend the night on it? If not, what will we do? Uncertainty is always present; you have to manage the consequences, limit the questioning chatter in your brain, order yourself to stay focused on not falling, keep your mind cool and shut out the panic that is always in the background, waiting to break out in an unguarded moment. When danger is a constant presence, remaining totally focused becomes a habit you can turn on. You enter a zone of great control and you feel an unwavering sense of purpose. You scan for opportunities and potential pitfalls. Uncertainty keeps you on constant alert. If all were certain, the very point of being here would be lost; there are no worthwhile frontiers when the ground and outcome are fully known.

Slowly and very painfully, the puzzle of the mountain unfolds in the struggle upwards. The fight is not with the mountain but against a struggling body and that part of you that wishes you were home, cuddled up on a warm sofa with a glass of wine. Comfort and certainty are never so appreciated as when you have the direct opposite, yet it is not hard to recall that not so long ago, while I was experiencing a period of comfort, I yearned for and dreamed up this very escapade.

But why, out of all the myriad options, had we chosen to come here in particular? Part of the answer is quite simple. I saw the mountain while trekking through the foothills and was drawn to the prominent and elegant line drawn by the sharp spur splitting the south face. There is an aesthetic in the routes that climbers choose – the most obvious and direct line is always favoured. Similarly, there is an aesthetic that applies to mountains – the more pointed and symmetrical, the more appealing they are to climb.

Although I could see almost the entire route from the sub-tropical languor of the densely settled foothills, it still projected a sense of wildness. It was apparent that the climb would start in the temperate rainforest at the mountain's base. From there, it rose in a series of vegetated buttresses to a wild, tumbling glacier, then continued up cliffs to meet the snowline, probably continuing along a curving, ice-fluted blade of a ridge that culminated in the rocky buttress we were on now. The route looked complex, challenging, full of variety and therefore tantalizing. Curiosity welled inside; I wanted to see what was up there. Even before the actual climbing, the approach through the thickly forested valley from the last habitation seemed adventurous. When I found out that it had repelled two other parties, the challenge became irresistible. My enthusiasm for the project made it easy to talk my climbing partners into joining the adventure.

The previous attempts on this side of the mountain didn't climb nearly as far as us. No human has ever been where we are now. Each view, each hand hold, each foothold is new. For the first time, the intricacies of this little corner of the planet are being revealed to a human brain, which, for all its frailties, must be the crowning glory of life on earth. But this conscious brain comes with baggage. Why? Why does this huge brain drive us to seemingly worthless ambitions, such as trying to reach a summit by one of the more difficult options – unimaginably hard to reach, painfully fought for, life threatening? And, if you do make it to the top, I have learned that part of you will wonder in astonishment that you suffered so much for a barren pinnacle of rock and snow.

We are the only organisms to pursue such goals. Animals do not climb high mountains or explore for any reason other than to fulfil their genetic or instinctive imperatives – to find food, to mate and sometimes to defend the territory that provides for them. This is not so for humans. We have a psychological, intellectually driven imperative over and above our instincts to find understanding of ourselves and to try to answer those ultimate questions: what are we doing here? Are we essentially good or bad? There is always great euphoria on reaching the summit. It is a vindication of one's worth (though, relative to one's whole life, it is somewhat momentary), but the summit invariably feels like a release from the struggle to prove oneself. The greater the struggle to reach the summit, the greater the euphoria. The feeling is so seductive that it becomes one of the principal motivations for future climbs. It is addictive.

We finally made it to the summit of Annapurna II, although the last section was the hardest of the whole ascent. The final 150 metres took us a day and a half to climb. We spent two nights on a flimsy 'sit-on-only' ledge hacked out of sugary ice stuck to the back of a chimney. So, when we stood on the summit, the euphoria was intense. But there was also a terrible sense of urgency as huge clouds amassed all around us. After a few brief hugs and back slaps, we hurried downwards to be engulfed in the fury of the storm. There were blinding flashes of lightning, the air buzzed with static, the wind literally knocked us over. Long, long into the night we finally reached sanctuary – our snow-cave a thousand metres below the summit – and fell asleep while the stove boiled the pot of water dry.

Our descent from that point should have taken us a day, but it took four. We struggled, in deep snow, facing avalanche danger, out of food and physically depleted. But we survived, and when we finally slithered down into the warm embrace of the villagers at Hoga, sated our ravenous appetites, savoured the sweetness of the oxygen-rich air and saw afresh the beauty of the life around us, we looked back and it all seemed worthwhile. Slowly and predictably, we began to dream of the next mountain.

Even though the physical traces of our ascent quickly vanished, the experience changed us for ever. We came away with an emboldened sense of confidence, a rich trove of experience and a far wiser appreciation of our own abilities and weaknesses. Because the whole project had been of our own making, with no external agency directing us to do anything, we had taken full ownership, which had allowed us to embark upon what was a big leap into the unknown. It gave us the impetus and resilience to overcome setback after setback. It unlocked hitherto unrealized potential in us. It certainly gave us the experience and confidence for our next 'big adventure': the first Australian attempt on Mount Everest.

Why not Everest?

It was not until I set eyes on Mount Everest that I began to ponder climbing it. Previously, it had held no interest for me because of the traditional style of climbing employed by every successful expedition – a large team of climbers using oxygen and supported by an army of high-altitude Sherpa porters who helped to establish a string of camps linked by fixed ropes up the mountain. But as I gazed up at the summit, I embarked on an idle daydream. I saw myself climbing up the western ridge and onto the easing snow-slope. Fancifully, I also began to think of an alternative approach. How wonderful it would be for a group of relative amateurs from the relatively flat continent of Australia to tackle Everest in a lightweight style – no oxygen cylinders, no huge logistical support. My dreams then went further as I contemplated pioneering a new route up the mountain.

Subsequent research revealed that the north face on the Tibetan (Chinese-controlled) side of Everest was the most feasible means of achieving that goal. Luck was on my side shortly after I began to entertain the idea when, on an expedition to the northern fringes of the Tibetan plateau, I managed to secure a permit for Everest from the Chinese government. So, a year after the Annapurna II climb, our small team flew to Lhasa. In 1984, central Lhasa still retained a medieval flavour, with hardly any sign of twentieth-century architecture. Pilgrims with long plaited hair, cloaked in homespun robes, some even in animal skins, perambulated around the holy Jhokang Temple. The thin air was pungent and sweet with burning juniper and highland rhododendron. It felt like the most exotic place on earth.

Seven hundred kilometres of gravel road westwards, the last fifty or so no more than a rough track, took us to the site of our base camp, itself twenty kilometres down the valley from the mountain. We arrived in the back of Red Army lorries – the most unlikely approach to a Himalayan peak I have ever had and a far cry from the overland approach of the pre-war expeditions. Back then, the mountaineers had started walking in the sub-tropical Tista valley in Sikkim before winding their way up into the cooler alpine heights, then crossing over high passes and onto the cold, desiccated Tibetan plateau – a journey of five weeks in total. In those days, it had been an expedition just to reach base camp. Since then, only the Chinese and a small number of recent foreign expeditions had been there. We arrived to find an empty, predominantly grey valley, the monsoon cloud hanging low over the rocky, desolate landscape. Everything was on a vast scale: the flat, braided river bed below the glacier was a kilometre and a half wide, and scree slopes rose thousands of metres into the clouds.

A few rusty cans and a couple of simple stone cairns bearing plaques in remembrance of vanished climbers – Mallory and Irvine, Tasker and Boardman – revealed expeditions had been there before. But now, except for the birds – mainly choughs, skylarks and ravens – the wild blue sheep, the Himalayan mouse hares and the marmots, we had the whole valley to ourselves. After setting up camp and establishing a route to the base of the mountain, we were committed to new territory on the north face. The question was: which way to go? Apart from what we could see before us, our only geographical references were a map drawn from the original 1921 survey and a grainy black-and-white photo from the Indian Air Force.

We saw no one else on the mountain and no further evidence of those previous expeditions. The climbing was challenging, the way ahead always uncertain, the weather frustrating – keeping us waiting for weeks at a time – and some of the dangers enormous, but with patience and caution, as well as some luck, we persevered. When two of us finally made it to the summit at sunset, eight weeks after reaching base camp, it felt like the wildest, most remote place on earth.

Less than 200 years ago, Everest was known only to local nomadic herders, some of whom knew that it was the tallest peak in the area (see Figure 8.2). However, it was only when Europeans came, equipped with sophisticated surveying instruments, that it gained the prestigious title of the highest mountain on earth. Decades later, British explorers turned their attention – and that of the world – to Everest. Their main goal was to climb to the summit, but they also explored the environs, mapped them and recorded flora and fauna. En route, they observed the Tibetans and their culture, which has proved enduringly fascinating to the West. There were eight attempts to climb the mountain in the first three decades of the twentieth century, if you include the first reconnaissance expedition of 1921.

After the Second World War, Nepal tentatively began to open its doors to foreign travellers, almost at the same time as the Chinese communists invaded Tibet, closing access to the north side of Everest. A possible route was found from the south side and, after two Swiss attempts, the British expedition of 1953 famously succeeded in putting two people on top. During the next

Figure 8.2 The north face of Everest, 1984
Photograph by Tim Macartney-Snape

decade, the Chinese succeeded in climbing the route pioneered by the pre-war British expeditions from the north but kept that side of the mountain closed until 1980. By the end of the twentieth century, all the main features had been climbed and every detail of the mountain was mapped and photographed. So now, on Everest, the age of physical exploration is over. However, it still offers a venue for personal exploration and it will always be a symbol of achievement.

I was on the south side in 1990, having walked to base camp from sea level on the Bay of Bengal, then continuing on to climb alone to the summit. At that time there were five other expeditions, numbering about forty climbers in total. Three of those went on to become big players in the first commercial expeditions to Everest a few years later. They, and others who saw the commercial potential of guiding paying customers up the world's highest mountain, began competing for business. It was a competition that indirectly led to the disastrous spring season of 1996, which witnessed eleven deaths, including two of the 1990 progenitors. The recounting of this tragedy in numerous magazine articles, several books (including a bestseller), an Imax movie and a full-length blockbuster only increased the mountain's allure.

Today, ascents of the mountain's two easiest routes with ropes, camps and supplies fixed in place by an army of local high-altitude porters are the norm and commonplace. During each climbing season, the Tibetan and Nepalese base camps can swell to over a thousand people, with hundreds venturing above. With improved cellphone and satellite communication, and the ready availability of meteorological data, there is comparatively little uncertainty or adventure left in climbing the mountain that way. Less than a decade after our relatively wild experience on the north face, the ascent of Everest had become a commodity – an experience that clients could essentially pick off the 'package-tour' shelf, with all services provided. They take no part in route finding, or the preparation of the route, or the establishment of camps, or any of the important decision-making. Others who know the route backwards do all of this for them. Climbing Everest has become 'touristy', and we are paying a heavy price in terms of overcrowding and the degradation of a once pristine mountain.

What has happened on Everest is a microcosm of what is happening everywhere in the travel and exploration industry. It could be said that mountaineering is a form of tourism, because it involves travelling to a place for recreation (see Chapters 10 and 11, this volume). I would argue that tourism shares some elements with mountaineering, just as it shares some elements with exploration, and they are all encompassed within the activity of travel. But we can be more definitive by describing tourism as a form of travel where the participant mainly engages in experiences – be they transport, accommodation, restaurants, tours and entertainment – created for travellers who are on holiday, whether for rest or a quest for adventure.

At one end of the tourism spectrum there is the 'self-guided' form, typified by buying the bare minimum of services, and at the other the 'package' form, whereby a whole tour is bought 'off the shelf'. Most travellers indulge in both forms to varying degrees. Relative to the average incomes of the participants, these services are cheaper than ever before because the costs are now calculated on the basis of big numbers. Mass tourism is in a positive feedback loop: the more people travel, the cheaper travel becomes. By 'positive', I mean it is self-fuelling: the end result may be economically positive for some of the local inhabitants, but it is invariably negative as it erodes the cultural and natural attributes that attract people in the first place. From being adventurous in the beginning, it is inevitable that the common travel experience becomes mass-produced and 'touristy'.

In an infinite world we might accept this trend as inevitable, but in the case of adventure tourism and exploration the rapidly diminishing resource of places untrammelled by millions of feet means we must try to manage the situation as best we can. The popular destinations in the Himalaya are reaching a tipping point beyond which they will start to suffer from the 'Everest syndrome' if better management is not implemented. Being unique, Everest has become an economic lodestone for those who can exploit it: from controlling governments that exact permit fees, to the locals who provide transport and porters, to lodge owners, shopkeepers and guides. Those who are lucky enough to reside within Everest's economic catchment area derive incomes that far exceed those of their compatriots who do not happen to live en route to the mountain.

As with any other iconic destination that has become a magnet for travellers, adopting a *laissez-faire* policy that allows the forces of commodification free rein is to invite despoliation and alienation. An alternative is to take control and manage the problem, so that the integrity – both natural and cultural – of these unique places is acknowledged and conserved to at least a semblance of its former self. Unfortunately, Nepal has a deeply ingrained political system that is characterized by a long tradition of nepotism, caste-based inequality and financial mismanagement. Despite government regulations on litter, enforcement is almost non-existent, leaving management of the issue to the guiding operators. Some strive to operate according to ethical 'best practice', but many, driven by economic motives, do almost nothing to ensure the conservation of the landscape. Concerned climbers have petitioned the government and suggested that the intense demand to climb Everest could be harnessed so that the enormous benefits of mountaineering tourism spreads to other parts of the Nepalese Himalaya, with consequent reduction of the pressure on Everest itself. So far, though, the government has failed to implement any policies in this regard.

In Tibet, the Chinese have adopted a pragmatic approach. The first step was to build a road to base camp for their first attempt on the mountain in the late 1950s. The undulating plateau terrain on the north side made this far easier than it would have been from the south, but it laid the foundation for what would happen next. Now, that road has been sealed all the way to base camp. Tourist buses stop only a few kilometres short of the camp, and a major resort complex – complete with mountaineering museum – is planned for a little further down the valley. On the mountain itself, the government has overseen the fixing of ropes and ladders all the way to the summit. At lower altitudes, rangers patrol the camps to assist climbers and ensure all litter is removed. So it is controlled and somewhat regimented in contrast to the Nepalese side.

In 1981, when I first visited the Solu Khumbu region of Nepal, where Everest is located, the trekking boom was still in its infancy. It would be more than ten years before commercial expeditions to the higher peaks became popular, so the experience of walking through those villages was quite different to what it is now. Though well-worn by trekkers even then, the paths, bridges, buildings and even the culture were still very traditional. It was authentic to its past. Wooden window shutters, rather than panes of glass, were the norm, and there were few corrugated-iron roofs. Almost all of the older people, and some of the youngsters, still wore the traditional style of dress. There were no mobile phones and no electricity, and if you ventured off the main trail you soon came upon people for whom a Western face was a novelty. Of course, change would have come, as it has almost everywhere, but the 'Everest factor' has enormously accelerated the pace of that change in Nepal. As the area's popularity and the number of visitors have grown, so too have the positive and negative effects.

Lodges have sprouted like fungi after warm rain. Electricity has arrived, so modern lighting and the constant flicker of televisions illuminate homes, exposing those locals who haven't travelled to the outside world. Such exposure is both good and bad. While it would be wrong to deny people the freedom to have what everyone living in the developed world has, the medium of TV undoubtedly homogenizes cultures. From having no telecommunication apart from high-frequency radio, the area has leapfrogged into high-tech culture with the rollout of cellular networks that give everyone instant access to the internet. This has hastened the erosion of local culture as tourists' posts on social networks promote the area as a 'must visit' destination. An interesting subject of study would be to assess whether sudden access to TV and the internet fuels the desire to possess modern goods and clothing and acts as an added impetus for people in remote areas to migrate to towns and cities.

The tourism boom and the building boom it generated have had an enormous impact on the environment. Large, colourful buildings now dominate villages that were previously more organically connected to the hillside in colours of stone, wood and whitewash. Voracious consumption of materials – mainly wood and stone – has depleted the area's forests, despite regulations against the cutting of timber. Due to the high cost of imported energy, large and unsustainable amounts of firewood are still burned in most trekking lodges. At higher altitudes, dried yak dung is used, depriving pastures and potato fields of valuable nutrients. Then there is the problem of litter. Traditional packaging simply biodegraded, so the locals' habit of casting it aside wasn't a problem. Meanwhile, most of the early expeditions' goods were packaged in reusable receptacles, such as tins and glass jars, rather than plastic. These were highly prized among the local villagers, who always put them to secondary use. However, the growth in plastic packaging and the rapid increase in visitor numbers have generated an unprecedented litter problem in the Himalaya. This is being slowly addressed as the local people begin to realize that a littered landscape is far from appealing to tourists. Human waste is another problem, especially at high altitudes, where the cold temperatures and almost desert-like dryness hamper the decomposition of faeces. Already, above base camp, the corridors up both of the regular routes are heavily contaminated with faecal matter.

In less than a hundred years, Everest and its environs have gone from wild and pristine, untouched by humans, to somewhere that is known in its entirety, with every detail mapped and photographed and the two easier routes festooned with old climbing debris, such as ropes, oxygen cylinders and tattered tents. Then there are the desiccated, frozen remains of those who have perished. Almost certainly, adventurous climbers lured by the prospect of ascending the world's highest mountain and lucrative sponsorship deals will keep coming to attempt the few unclimbed lines that remain. Increasingly, though, they will feel as if they have entered a giant extreme adventure park. The sense of wildness that so attracted us to Annapurna II and the north face of Everest has entirely vanished.

The fully guided, fully catered experience eliminates most of the decision-making, most of the risk and most of the adventure. Many of the new breed of 'list-tickers' have little interest in lower, lesser-known and less-visited mountains or in the richness of the full mountaineering experience. This 'no holds barred' approach indicates a sad loss of dimension in the mountaineering experience. It has mostly disappeared from the other 8,000-metre peaks, too. Tagging summits – or peak-bagging syndrome – has become a fashionable pastime that guiding companies are only too happy to facilitate. They offer complete package tours throughout the Himalaya and so, one by one, all of the region's mountains are succumbing to commodification and its trickle-down effects.

This form of 'mountaineering' has become just another tourism product, albeit an extreme one, but it is attracting a growing number of mountain tourists. It could be seen as an extreme version of the classic mode of mass tourism – the bus tour. This is certainly convenient for the traveller and profitable for the operator – hence its popularity – but it has the great disadvantage that it places a barrier between the traveller and what they want to experience. The ease of having everything organized carries the penalty of a diminished experience. Moreover, the large numbers of people who now visit places with no sanitary or litter-disposal facilities are contributing to the ruination of once pristine environments. The physical, aesthetic and reputational effect on traditional climbing and trekking values has been profound. They are the only truly authentic exploratory experiences left in the world's greatest mountain range, but our appreciation and regard for them is being eroded and undermined by the 'Everest syndrome' that is spreading, unchecked, across the range.

However, it is never too late to instigate better management. And it should be remembered that the Himalaya is vast. Away from the narrow, easiest routes to the base camps, and certainly on the approaches to the summits of the fourteen 8,000-metre mountains, most of the Greater Himalaya remains unspoiled by the excesses of tourism. It is still possible to trek and climb in literally hundreds of valleys across the region and be the only humans for miles around.

Nevertheless, there are big challenges with respect to maintaining a balance between the local people's need for infrastructure development, the sustainable management of the region's pastureland, timber and herbal medicines, and the preservation of the Himalaya's natural integrity. Moreover, all of these challenges are now inextricably enmeshed with tourism. This growing resource has the potential to contribute significantly to local economies. Tourism that preserves a relatively pristine environment around the earth's highest mountains will ensure that the world keeps coming, and sustainability is strongly encouraged if the principles of responsible tourism are followed. Travelling to a place in a manner that influences change for the better can also be a motivation to visit that place rather than somewhere else. Adventurous travel will have reached maturity when those who do it, wherever they go, strive to do it sustainably.

Part III

Adventure

9 Free-falling the water column

Raptures and ruptures of the deep

Felicity Picken

Introduction

Very few places provoke a sense of exploration and adventure like undersea does. As a frontier that remains as mysterious to humans as outer space, undersea worlds have a long association with extreme experiences. The threshold of this extreme comes into being at the point of immersion, disturbing the material realities of land where the body, mind and emotions become conjoined in 'new protocols of relating' (Pell 2012). Here, we discover a partially 'other self' that results from transgressing the familiarity of land for the deeply unfamiliar conditioning of life beneath the waves.

Within this domain, very few practices engage with the transgressive properties of being undersea more than free diving. Supporting one of the most intimate relations with oceanic space, the free diver submits to a loose 'choreography' (Pickering 1999) of free fall in an ocean column. Testing the limits of human *being* undersea, without the technological assistance of SCUBA, divers rely instead upon negotiating an altered state, of blood gases, cardiac function and metabolism. These negotiations make free divers important sites for transforming medical knowledge, and also for transforming our sense of the relational core of being, thinking and feeling.

In following these lines of exploration, this chapter follows the work of others who are interested in better understanding states of 'being aqueous' (Pell 2012) and 'becoming aquatic' (Pell 2014). The chapter gains access to the 'lived' or experienced dimension of transformation in the oceanic column through free diving as a *frontier activity* that provides an opportune 'threshold of adaptation' (Pell 2014: 107) beyond which normative boundaries of human experience begin to shift. The specific focus on the experience of narcosis is an opportunity to consider a rupture in the orderly world of free diving, just as free diving itself is a rupture in the orderly world of life on land.

The chapter addresses the altered state of narcosis – or 'rapture of the deep' – by outlining how it is explained, how it is evoked and how it is negotiated. First, the discourse of narcosis authorizes what counts as an explanation for narcosis, conceived mainly for its pathological qualities in endangering human life. Second, the way that narcosis is evoked as a creative expression is explored through the acclaimed short film *Narcose* (Gautier 2013), directed by Julie Gautier and featuring champion free diver Guillaume Néry. Third, online accounts of free divers' experiences of narcosis are mined for a fuller (mind, body, emotion) picture of the experience, focusing on how some divers negotiate the experience of narcosis by letting it invade them. These accounts provide some of the best evidence of how narcosis is *lived*, how it engages affective, physical and cognitive states of being and how the diver is transformed through the free diving experience of narcosis.

Each of these sections will contribute to the question of how the radical externality of the ocean depths imprints human bodies, minds and emotional states through journeys of falling to depth. First, though, the chapter will explore the ocean as an ongoing frontier for predominantly land-dwelling humans before discussing the efficacy of free diving as a probe into the nature of intimate oceanic frontier experiences.

The deep as frontier

The ocean deep has been an ambiguously situated frontier, perceived to be simultaneously an outer limit to human life and a condition of it. This is evident in Strabo's philosophical prose, written more than 2,000 years ago:

> We are in a certain sense amphibious, not exclusively connected with the land, but with the sea as well ... The sea and the land in which we dwell furnish theatres for action, limited for limited actions and vast for grander deeds.
>
> (Cited in Steinberg 1999b: 368)

And similar sentiments have been expressed more recently, in the reinvigorated language of twenty-first-century geographers of the sea:

> By wayfaring once again – much like born again children, with the same curious disposition toward wonder, with the same thirst for astonishment – we return to a world that is continuously unfolding, brimming with life, dripping with complex but elegantly simple relations, subject to endless flow, transformation, and change.
>
> (Vannini and Taggart 2014: 100)

As humans shift beyond their exclusive land-based habitats and enter these spaces of comparative 'endless flow, transformation, and change', they 'extend the ecological limits of humanity' (Peters 2010: 1262). In doing so, the usual, land-based realities that have secured and shaped the many experiences of being human are exchanged for the radically different realities of undersea – a place where humans enter into a highly experimental relationship with their surroundings. This highlights both the significant imprint of the mundane conditions of land upon human *beingness* and the potential for new experiences of being human, beneath the sea (Picken and Ferguson 2014). While inevitably directed towards ontological questions, such as the limits of being human, the implications of these new experiences are increasingly pursued as empirical questions across the social sciences and humanities. Collectively, each point of immersion increasingly shapes an age in which undersea worlds are entering more deeply into human culture and history (Serres 1995; Mentz 2009).

What are recognized today as the marine and oceanic sciences are traceable to fifteenth-century ichthyologists (who studied fish), through to the 'voyages of discovery' of eighteenth- and nineteenth-century maritime activity (Ellis 2005) and the formalization of the marine sciences in the mid-twentieth century (Rozwadowski 2005). The transformative properties of the oceans, as a consequence, were extended into scientific practice itself and particularly to deep sea science, which, at the time, was literally blind: it relied instead on *feeling the way*, listening, trial and error (Ellis 2005; Rozwadowski 2005, 2010). Gaining access to this 'sealed volume' of knowledge and instruction (Matthew Fontaine Maury, cited in Rozwadowski 2005: 67), which did not record or respond to activity in the same way as land, required new technologies of observing, sampling, mapping and 'doing' knowledge. As Hamilton-Paterson (2011: 1) notes: 'Until a certain moment in history there must have been a conceptual impossibility in the idea of a sea chart without a coastline on it.' Practices that had become commonplace on land were all frontiers in the equally ambitious desire to know the planet as an entity that is greater than land. Within this context, as an extreme and vast environment, oceans were considered to be as 'foreign to our understanding as another universe' (Ellis 2005: xi), and through these absences of knowledge, 'the deep [is] a site where the imagination has had full rein' (Farquharson and Clark 2014: 2).

The sea has a long history of 'taboos and rituals' (Peters 2010: 1266). The likes of sea monsters and lost treasure have served as points of access to underwater worlds, acting as gatekeepers that either lure or repel with the promise of bounty or retribution (Picken and Ferguson 2014). These amount to early domestications of oceanic spaces in stories, beliefs and rituals that perform them as a 'cultural environment' (Edwin L. Hutchins, cited in Lambert *et al.* 2006: 483). Today, the visibility of undersea worlds in stories, images and film is enhanced by global media and commodification, but still characterized by the *invisibility* of undersea, where the imagination completes what the eye cannot see. While sight becomes continually perfected – which is the aim of the scientific endeavour – the unseen continues to be filled in and, arguably, refashioned and adulterated by the imaginative resources of our time. The power of these imaginative resources, including powerful literary archetypes (Farquharson and Clark 2014), is a significant shaper of knowing, and therefore relating to, oceanic worlds. The blending of fact and fiction, science and imaginary still characterizes many of our oceanic relations. The coupling is particularly obvious in the growing presence of oceans in the lives of more people for sport, at leisure and at play.

The global, twenty-first century has witnessed an accelerated consumption of oceanic worlds as experiences, where oceans are encountered for purposes that meet socially derived desires like pleasure, self-fulfilment and interest (Cater 2010). For tertiary and quaternary 'experience economies' (Pine and Gilmore 1998), the new economic value is dependent on keeping these worlds intact. Oceanic relations are performed through the creative, media, cultural, leisure, tourism, sport and entertainment industries, facilitating greater access for larger populations and among the most significant conduits for encountering oceanic worlds (Mentz 2009).

Framing these is 'the blue planet', a term that has gained widespread familiarity and acknowledgement of the oceanic nature of the Earth. The image of a distinctly blue sphere, first seen from space in the 1960s, exposes the seas and oceans as our most distinguishing and dominant marker in the solar system, and the most iconic symbol of our age (Sachs 1999). That said, while in the twenty-first century many people know that we live on the blue planet, we are only just beginning to comprehend what this means. Calls for a better understanding of 'blue' culture (Mentz 2009), urbanism (Beatley 2014), leisure (Merchant 2011), geography (Peters 2010; Steinberg 1999a, 1999c, 2001, 2013), history (Rozwadowski 2010) and social anthropology (Probyn 2015; Helmreich 2009) are beginning to produce a sizeable body of knowledge. This scholarship is both empirically grounded and theoretically mindful that, even as oceans are becoming more socialized, they remain profoundly 'other' to social life (Helmreich 2009). Human beingness, even on the blue planet, is primarily a land-based affair.

Crossing the threshold between land and ocean are a multitude of relationships that testify to the various ways through which humans are coming to terms with life on a blue planet. This equates to a greater acquaintanceship with the oceanic frontier through greater access, variety of encounters and knowledge, permitting greater permeability between land, air and oceans. Performing both a natural and a cultural '*reversal* between the things of the land and the things of the sea' (Mack 2011: 188; emphasis added), as humans assimilate more of the blue planet into their consciousness and material lives, as they engage with these places more intensely and frequently, a fundamentally different set of planetary realities compels new challenges and possibilities. This has implications for the nature of human experiences and conceptions about the rightful place of humans whose confinement to land, in the face of this vaster frontier, re-emphasizes a subject who is an *earth*- more than an *Earth*-based being (Wigen 2006; Vink 2007). Oceanic realities are 'inherently novel' (Pell and Mueller 2013), deviating from, and therefore emphasizing, the largely invisible, routine stability upon which land-based life is enacted and written. They provide an opportunity to increase our understanding of how these *material conditions* – across land and oceanic spaces – imprint all human action and are therefore inherent to all human life. There is still much to be discovered within the oceanic frontier, within the presence of an expanded sensorium that is both of this world and alien to that part of the world with which we are most familiar.

Not surprisingly, undersea has been likened to a nearer outer space (Farquharson and Clark 2014). It is the planet's three-dimensional *inner-space* equivalent (Ellis 2005), where 'elemental forces', or non-humans, 'thrash it out among themselves' (Clark 2011: xiii). Both spaces are vaster than land, far less understood and, paradoxically, inhospitable to the immediate experience of human life, yet ultimately vital to it. While they seem friendlier to humans at a distance, part of the recent affinity with oceans is the growing expert and popular consensus about our proximity to them, in the interconnectedness between our lives and the oceanic nature of being on the planet (Beatley 2014). Within this enlivened context of interdependence, one of the most common depictions of oceanic spaces revolves around their transformative potential. As *transitional* and *transgressive* places, they provide 'a perennial resource of escape and transformation' (Hoare 2014: 17) and are, not insignificantly, aligned with contemporary metaphors of liquidity and flow that accompany the accelerated upscaling of social, economic and environmental units of analysis to the level of the global (Bauman 2000; Castells 2006; Helmreich 2011). At this emerging planetary scale, the ocean looms much larger than it did when the world was more fragmented into islands of continents or regions of land. As life increasingly takes place within a global sphere, so it takes place within an oceanic place: 'Our world is a water world. The oceans and seas are entwined, often invisibly but nonetheless importantly, with our everyday lives' (Anderson and Peters 2014: 3).

Against this background, this chapter considers the watery world of free divers and how they succumb to – as part of negotiating – a state of narcosis. Those who have witnessed, experienced and recorded their recollections of the rapture produced by falling to depth are drawn into a discussion about the transformative potential of oceans to the range and depth of human consciousness and experience.

Falling to depth

Early marine scientists in the nineteenth century believed that drowned people would fall to a depth in the ocean at which their density would equal that of the water and they would thus 'find their depth' (Rozwadowski 2005). The Azoic Hypothesis insisted that all life would fall to this degree zero within the oceanic depths and that this would occur well before its maximum on the sea floor (Grant 2014). The deep itself was a dead zone since high densities, low temperatures and the absence of light all worked against the prospect of deep sea life, but as explorations travelled deeper, the depth at which life was possible continued to be revised (Rozwadowski 2005). Disproven, league by league, each claim made way for more accurate, more remarkable and sometimes greater truths. This pattern of revision, remarkableness and discovery is still evident today (Ellis 2005, as with free divers who have consistently extended the limits of depths thought possible by medical science (Ecott 2001).

Free diving is one of the oldest human engagements with undersea worlds and an activity that humans share with other aquatic life, including penguins, seals, whales and dolphins. It began, it is logical to suppose, with a desire to exploit the ocean's generous bounty, but some have suggested that the earliest free divers were pivotal to the process of human evolution. Westenhöffer proposed Aquatic Ape Theory in 1943, and it was subsequently revised by Hardy in 1960 and Morgan in 1972 as part of the scientific study of human evolution. It counters Savannah Theory, which contends that humans evolved in adaptation to the specific conditions of savannah environments, by suggesting that aquatic environments played a greater role. The aquatic ancestor hypothesis is based on the fact that many human attributes are not shared by other land-dwelling species, but are held in common with aquatic creatures. The notion is that an ancestral ape inhabited a semi-aquatic (terraqueous) environment and developed 'in-built aquatic adaptations' that substantially transformed and distinguished it from land-dwelling apes (Morgan 1989: 22). These adaptations include the loss of body hair, changes to subcutaneous fat distribution, the shedding of tears, bipedalism as an outcome of muscular–skeletal changes facilitated by buoyancy and wading, and the ability to hold breath underwater through successive attempts to dive. An aquatic phase in human evolution is one explanation for our comparative 'increase in thoracic innervation' (MacLarnen and Hewitt 1999), and thus our ability to undertake *voluntary breathing* (Verhaegen 1985), including holding our breath.

One of the main obstacles to proving (or disproving) the Aquatic Ape Theory is that it proposes a history with a distinct lack of evidence (Jameson 1994). The lack of a fossil record maintains the theoretical status of the aquatic ape with questions of whether such a record exists and whether it will ever be recognizable or recoverable from the oceans. Here, again, is a confrontation with the transformative capacity of oceanic worlds, in this case on the production of scientific knowledge in confronting what Szerszynski (2016: 618) terms the 'fluid memory' of oceans, where 'the motion of flows, eddies or vortices [is] intrinsically more forgetful' than on land. The dive reflex is another way of describing some of the physical and behavioural traits that humans share with aquatic life. It is described as an evolutionary adaptation and is evident in the way the body responds to immersion, including slower heart rate and the reduction of blood flow to muscles and extremities (Ecott 2001). Even as these theories remain theoretical, they describe observable and distinctly aquatic capacities that differentiate humans from similar species on land as well as a long, enduring and special relationship between humans and free diving.

The modern form of free diving – as both a competitive sport and a recreational pastime – has followed the growth in recreational SCUBA diving since the 1960s and represents a return to 'naked diving' but at a 'new level'. Like SCUBA diving, it has attracted significant levels of participation, having first gained popularity along the Mediterranean coasts (Ecott 2001). It has been described as appealing to the 'spirit of the age', as 'a new sport for a new century', and as the 'Zen of diving', with its ideal of tranquil and meditative states, and because it is imbued with poetics, other-worldliness and mysticism (Ecott 2001: 322). In the competitive world, it is also known as *apnoea* – a Greek term meaning 'without breath' (Ecott 2001: 298). The training for free diving is an adaptation process that involves *reacquainting* the adult body with the infant practice of using the lower abdomen and lungs for breathing. As one diver recalls: 'Rediscovering our breath can be one of the most difficult things to do. It involves letting go of tension, doubts, fears, old patterns, assumptions and expectations' (Farrell 2006: 22). The diver must also learn to lower their metabolism and enter a state of deep relaxation to preserve as much energy and air as possible. Among the free diving community, this is regarded as the 'purest version of the sport, and of man's [*sic*] relationship with the sea' (Ecott 2001: 300). Rupturing this ideal state of tranquil, free-falling the water column is the experience of narcosis.

Explaining narcosis

Narcosis occurs in both SCUBA and free divers. In its simplest form, it is the result of a 'gaseous attack on the central nervous system' that happens under pressure of depth (Ecott 2001). However, the effects are difficult to measure since depend on multiple variables of time, place, mind, mood and body (Gibbs 2017). Therefore, while its cause is understood and there are medical procedures for reducing its effects, knowledge about what these effects will be and when they will occur is inexact and imprecise. Hence, it could be said that each experience of narcosis is unique and each telling of those experiences adds to the shape of what it is and does. Narcosis occurs within a body that is already under significant pressures that require various responses across the respiratory and cardiac circuits. The lungs, spleen, heart and major blood vessels are all compressed by the increased pressure of depth (Ecott 2001). Medically, narcosis is an anomaly to be managed, if not avoided, for the very real dangers it poses to human life both during the dive (disorientation leading to distress or mortality) and after the dive (with higher concentrations of partialized air within the circulatory system).

Symptoms that are similar to intoxication have been described by divers since the early 1800s, with science finally identifying the cause as nitrogen in the 1930s (Dovala 2015). They are emphasized by the language used by divers, such as 'narked', 'drunk' or, in Jacques Cousteau's enduring phrase, the 'rapture of the deep':

> I am anxious about the line but I feel really wonderful. I have a queer feeling of beatitude. I am drunk and carefree. My ears buzz and my mouth tastes bitter. The current staggers me as though I had had too many drinks.
>
> (Cousteau 1988: 30)

Similarly, the 'martini rule' is an inexact colloquial term that divers use to describe depth intervals at which they will become affected by narcosis, each registering on the body the equivalent of drinking one martini (McLaren 2017). Impaired mental ability, including reaction time, orientation and decision-making, has also been likened to a form of delirium or early anaesthesia, creating what is referred to as a 'reversible alteration in consciousness' (Löfdahl *et al.* 2013: 17). That said, reports of 'such a feeling of happiness' at depth can be so intense that they tempt the diver to stay there (Ecott 2001: 306).

In addition to intoxication, the sensation produced by narcosis has been described as both 'euphoria' (a positive or appetitive emotional state) and distressing or confusing (defensive, negative emotional states) (Löfdahl *et al.* 2013). However, like the diver's cognitive processes, emotions are poor guides to action. An additional danger that is present in diving narcosis is that negative emotional content, such as feeling disoriented, can be experienced as a positive emotion. The sense in which narcosis can be an enjoyable and even fun experience, like attending a cocktail party beneath the sea, is met with the dangers of this altered state of consciousness that is exacerbated beneath the sea. Much of the research about narcosis in divers is dedicated to reducing acute distress or mortality by managing, controlling or mitigating the risks during dives and improving medical responses post-dive (Löfdahl *et al.* 2013). To this end, various safety measures are built into responsible diving practices, but the nature of the condition and the environment in which it occurs presents a significant challenge. Key among the risks is the diver's own ability to recognize and communicate the symptoms of early narcosis. This is undermined by the fact that one of the first symptoms of the condition is the impairment of the diver's ability to understand they are experiencing it. Narcosis, to the affected, as with other mind-altered states, is experienced almost completely without an 'outside' that can recognize it.

Evoking narcosis

The cognitive and physiological aspects of narcosis are better understood than the emotional states that are produced. There is a solid bank of knowledge about what narcosis is, how it imprints the body physically and psychologically and how to recognize these imprints. Less, however, is known about how it *feels*, although the field of Critical Medical Humanities has at least explained why this matters (MacNoughton and Carol 2016). Knowledge about the affective dimension of narcosis can be found amongst dive communities where creative, written and oral testimonies are expressed and shared. Many of the recollections and testimonies of free divers appeal to the felt or emotional response. This continues a tradition established by early divers, including the famous French explorer Jacques Cousteau, who, through his own recollections and memoirs, coined the now widely recognized term 'rapture of the deep' to describe his experience of narcosis. In the word 'rapture' there is a denial of a simplified representation or explanation of the experience, and this multidimensionality is embraced by filmmaker Julie Gautier in her documentary *Narcose* (2013).

This fourteen-minute film documents the effects of narcosis through the recollections of Gautier's husband, the renowned free diver Guillaume Néry. A follow-up to the same team's *Free Fall* (Néry and Gautier 2010), it adheres to the tradition of an earlier acclaimed free diving film, *Le Grand Bleu* (*The Big Blue*; Besson 1988), in its attention to style, aestheticism and spectacles of visual pleasure (Beugnet 2004). *Narcose* documents Néry free diving into narcosis, taking poetic licence to convey the thoughts and feelings the condition engenders. Following his preparation for the dive, then descent along the line to depth, he crosses the point at which narcosis begins, and the journey becomes an equally interior one, recalling the cognitive and emotional content of Néry's experiences.

At one point in his journey, Néry finds himself running across the ocean floor as if fleeing from a pursuer. His gives the impression of being immersed in a gelatinous substance, with the water retarding his efforts, which are disproportionate to their effect. The weight and friction of the water put him into what the earth-based human interprets as slow motion. His body-in-immersion resists his panic-stricken will and desire to run. The heightened flight response returns an inadequate physical outcome – conveying entrapment within the ocean depths, but not drowning. The dramatic countermovements highlight the normative logic that is a consequence of land-based thinking, since slow-motion running across the ocean floor is not an anomalous logic *beneath the sea*. In this way, the anxiety of Néry's recollection is produced in the relationality of the narcosis experience, of the inadequacy of his land-based being to negotiate new oceanic realities.

It is through the mechanism of this interior journey of ascent and narcosis that the film shifts from realism to surrealism, hallucinations and dream sequencing, which describe the visions, thoughts and anxieties experienced by a diver's collective experiences of being 'narked'. It is a film where actual and imagined, science and fiction, nightmare and dream become entangled, imprinted on the body, mind, emotions and memories of falling through the ocean column with one breath. Through a journey of ascent with a trail of 'hallucinations', the film exposes the viewer to sensations that are variously pleasant, peaceful, erotic, frightening and anxiety-ridden; exchanging dreamlike and nightmarish realities. Not unlike those literary and filmic characterizations of 'near-death' experiences, the film is both surreal and raw in *presenting* the emotional, cognitive and even spiritual journey of narcosis. In doing this, it exposes us to a less technical but more 'lived' (visceral, affective) appreciation of the experience of falling to depth – and into narcosis – in the ocean column.

Evoked knowledge is not the kind that stands up to much epistemological rigour, yet it captures the interstitial, precarious and imprecise nature of narcosis at the edge of reason, feeling and life. While the film makes an important contribution to evoking narcosis through abstract – and visually enchanting – expression, actual experiences vary according to the circumstances of each dive. Narcosis occurs at unpredictable depths and plays out differently within individuals at each time and in each place. In emphasizing this, individual testimonies of divers' recollections of narcosis add an important, subtler layer to our understanding of the transformative capacity of the undersea frontier. There are also recollections that point to shared or similar experiences among members of the diving community, for example in divers' explanations of how they negotiate, go along with or 'play' narcosis in order to live through it.

Negotiating narcosis

In an entry posted on the popular dive site *Deeper Blue*, free diver Eric Fattah describes his sense of 'giving in to' or allowing his 'self' to become merged with the performance of his body's motion through the saltwater depths: 'Soon, I was sinking past 20m into the blackness, with no bottom in sight. I pointed my light into the darkness and tried to sleep as I sank. Down and down I sank, and I felt the narcosis begin' (Fattah 2003). In this case, the merger takes place through the diver *giving in to* passive sinking and allowing himself – or 'trying' – to fall asleep. This description corresponds to veteran diver Tarek Omar's observations of narcosis in action, and what he describes as the diver falling 'under the control of the sea' (cited in Bower 2017). The 'sea', in this case, is constituted by all that the diver (his physical, psychological and emotional processes) is enmeshed within (saltwater, pressure, darkness and falling).

In another discussion on *Deeper Blue*, one contributor ('tcrusson') suggests, 'if you let narcosis invade you, it can lead you to a real state of bliss!' This seems at odds with general 'good sense', which stresses narcosis as a condition to be expelled from the body, defeated by the mind, resisted, avoided or at least managed. Indeed, this free diver and others who join the discussion are well aware of the threat narcosis poses:

> 'tcrusson': Distortion of time … Changes in the perception of colors … one's sense of logic is seriously hindered. It can feel really weird to experience these effects, and if you try to fight them, well, you can't, so you can easily freak out and go into full panic mode.
>
> 'Jon': Of course the most dangerous part of deep air diving is not KNOWING you're narked until it's too late.
>
> 'Pekka': Everything seems 'wavy', everything sounds distant, you can't think clearly, you feel a buzzing all over your body, feeling 'drunk', can't concentrate on kicking, confusion, fuzzy vision, fear, dizziness.
>
> 'Lynn': The nitrogen hampers or blocks proper impulse transmission in the central nerve system. Result: some sort of drunkenness varying from narrowed consciousness to euphoria, primal fear or hallucinations

> (a buddy of mine saw her children sitting in front of the TV at 40m of depth!).
>
> (*Deeper Blue* 2001)

A diver who is *invaded – taken over –* by the ocean is (sometimes dangerously) *open to being open to* the other. This other may be the abstract 'transformative process' of narcosis, or the sea and its effects, or the body itself. One important part of the process of surviving the effects of narcosis is to remain 'in tune with the silent messages and sensations that the body uses to signal its state' (Kurra *et al.* 2015: 15). What Kurra *et al.* (2015) describe as 'learning to be affected' by the undersea world, and learning to register this within the body, is part of what might be called a *healthy invasion process.*

Similar conclusions have been made with respect to SCUBA diving, where, on discovering a partially 'other world' (beneath the sea), the diver becomes a partially 'other self'. In both cases, the diver must learn to give in to the dive and entrust their self *into* the world, since the journey and return are dependent upon a process that involves 'assimilating the radical exteriority of a mind, body and ocean so that there is only then "indeterminate, partial exteriority"' (Picken and Ferguson 2014: 335). This involves becoming attuned to the oceanic atmosphere and, in SCUBA diving, the equipment through which breathing is re-circuited. For free divers, it is the act of falling, of going along with a descent of heightening darkness, density and compression that issues the act of trust and ignites compensatory reorganization of vital life systems, thoughts and emotions as the water column shifts across the self.

Conclusion: returning to land

In the comfortable space of land, where chapters like this one are written, our language is increasingly suffused with terms like 'hybridity', 'relational entities' and 'post-humanism'. These testify to growing recognition of the porosity between the self and the 'other', of transgression and transformation as the ordinary state of being. These, theoretically, make frontier spaces of all of us. In the less comfortable space of undersea, where frontier realities get up-close and personal, where the mundane conditions of life are exchanged for strange, non-terrestrial ones, the free diver who is immersed but also somewhat suspended, and infinitely small, demonstrates what this means.

For most of human existence, with the exception of our special ability to hold our breath, the experience of being human has been that of a land-dwelling creature with a body that requires air. One of the consequences of this commonplace fact is that human experience has almost exclusively been composed without a consciousness of life beneath the sea (Picken and Ferguson 2014). As this begins to change, as human gain greater access to oceanic realities, increasing opportunities to understand the implications of this expanded ecological limit have followed.

This chapter has explored some of the embodied, cognitive and emotional transformations that take place when divers free-fall the oceanic column. Particular attention was paid to one of the most dramatic measures of this transformation in the altered state of narcosis or 'rapture of the deep' (Cousteau 1988). Drawing upon experiences of narcosis as they are expressed in medical and scientific discourse, on film and in testimonies from within the free diving community, the chapter explored one way through which the oceanic frontier extends the 'range of humanness [that is] possible in our era' (Brasher 1996: 815).

The state of narcosis was used as a provocateur for understanding some of the ways in which the anomaly of free-falling the water column imprints the body, mind and emotions and thinking about the adaptive potential of frontier spaces (species) within this. The free diver not only submits to the deep oceanic frontier but in doing so becomes part of it, as a *frontier being*, continuously testing the limits of a single breath in an oceanic column. The frontier being is perhaps best conceptualized after Latour's (2004) suggestion that the body is an 'interface'. He describes a self 'that becomes more and more describable when it learns to be affected by many more elements ... in rendering the body sensitive to what these other elements are' (Latour 2004: 206). This affords the free diver a special imprint insofar as it is an experience that is novel and not widely shared – being manipulated through the water column, and perhaps through narcosis, to become affected in a way that we, as a species, are only beginning to describe.

The free diver, negotiating experiences of narcosis, exposes how the self is always a *convertible self* (Margolin 1995), who, according to Franklin (2002: 193) 'calls up' and 'communicates' with the world 'through specific embodied configurations, postures, exercises, stances, mediations and observations'. More than this, some studies of immersion of the body within 'the body of water' have suggested that there follows a state of indistinguishable presence (Pell 2014: 106). This echoes Probyn's (2015: 78) discovery that such oceanic encounters produce a heightened awareness of what is present elsewhere but less noticeable – 'the durability of conjoined life'. Free-falling to depth allows a renegotiation with the world and within the body, which becomes exposed to new kinds of sensations (Rozwadowski 2005).

The return journey to depth produces a mobile body that is, following Szerszynski's (2016: 614) generous definition, an emergent phenomenon, an expression not only of a body through space, but of reciprocal relations of momentary, planetary self-organization. Clark (2011: xiv), equally mindful of the sociability of planetary acts, describes a similar kind of self: 'As fleshy, sensuous creatures, we have always been exposed to the energy and the inertia, the flow and the congealing, the mobilization and the halting of the earth.' Beneath the sea, we are 'viscerally out of our domain'; when we enter blue spaces we are *green* (Probyn 2011: 2). In this way, immersion provides an 'exceptional means of surrendering the ego' (Pell 2014: 102); and in this green surrender, which is common to the feeling of intoxication described by divers, is a willingness by some to become acquiescent to the ocean, the column and narcosis. At the same time, the idea that some divers negotiate openly with narcosis and sometimes invite it highlights, somewhat perversely, that there is a vital distinction to be made between narcosis as a tale to be told, a personal experience to be relived and expressed, and narcosis as the impersonal statement of cause of death. While it is not death that free divers wish to invite, the thrill of the edge or frontier brings an experience that they know can take them *close* to it. Jacques Cousteau (1988: 32) described the dangerous distinction perfectly when he admitted to being 'receptive' to narcosis, explaining, 'I like it and fear it like doom.'

References

Anderson, J. and Peterson, K. (2014) 'A Perfect and Absolute Blank: Human Geographies of Water Worlds'. In: J. Anderson and K. Peterson (eds) *Waterworlds: Human Geographies of the Ocean*. Farnham: Ashgate, pp. 3–19.

Bauman, Z. (2000) *Liquid Modernity*. Cambridge: Polity Press.

Beatley, T. (2014) *Blue Urbanism: Exploring Connections between Cities and Oceans*. Washington, DC: Island Press.

Besson, L. (director) (1988) *Le Grand Bleu*. Weintraub Entertainment Group.

Beugnet, M. (2004) 'French Cinema of the Margins'. In: E. Ezra (ed.) *European Cinema*. Oxford: Oxford University Press, pp. 283–298.

Bower, E. (2017) 'Top Diver's Death Casts Long Shadow over Deep Beauty of the Blue Hole', *Observer*, 27 August. Online. Available at: www.theguardian.com/world/2017/aug/26/blue-hole-red-sea-diver-death-stephen-keenan-dahab-egypt. Accessed 29 August 2017.

Brasher, B. (1996). 'Thoughts on the Status of the Cyborg: On Technological Socialization and Its Link to the Religious Function of Popular Culture', *Journal of the American Academy of Religion*, 64(4): 809–830.

Castells, M. (2006) *The Theory of the Network Society*. Cornwall: MPG Books.

Cater, C. (2010) 'Any Closer and You'd Be Lunch! Interspecies Interactions as Nature Tourism at Marine Aquaria', *Journal of Ecotourism*, 9(2): 133–148.

Clark, N. (2011) *Inhuman Nature: Sociable Life on a Dynamic Planet*. London: Sage.

Cousteau, J. (1988) *Silent World*. London: Penguin.

Deeper Blue (2001) 'Nitrogen Narcosis', 6 December. Online. Available at: https://forums.deeperblue.com/threads/nitrogen-narcosis.19695/. Accessed 17 March 2018.

Dovala, J. (2015) 'Rapture of the Deep: The Peculiar and Subtle Ways Nitrogen Can Influence Our Behavior Underwater', *California Diver*, 3 February. Online. Available at: http://californiadiver.com/rapture-of-the-deep0204/. Accessed 15 November 2017.

Ecott, T. (2001) *Neutral Buoyancy: Adventures in a Liquid World*. New York: Grove Press.

Ellis, R. (2005) *Singing Whales and Flying Squid: The Discovery of Marine Life*. Guilford, CT: The Lyons Press.

Farquharson, A. and Clark, M. (2014) 'Aquatopia: The Imaginary of the Ocean Deep'. In: A. Farquharson and M. Clark (eds) *Aquatopia: The Imaginary of the Deep*. London: Tate Publishing, pp. 2–13.

Farrell, E. (2006) *One Breath: A Reflection on Freediving*. Devon: Pynto.

Fattah, E. (2003) 'Leaving the World Behind II: Return to Tahsis', *Deeper Blue*, 3 January. Online. Available at: https://forums.deeperblue.com/threads/leaving-the-world-behind-ii-return-to-tahsis.31376/. Accessed 17 March 2018.

Franklin, A. (2002) *Nature and Social Theory*. London: Sage.

Gautier, J. (director) (2013) *Narcose*. BlueNery Productions, Saint Thomas Productions and Forêt Bleue. Online. Available at: https://vimeo.com/95182734. Accessed 25 November 2017.

Gibbs, N. (2017) 'What is Nitrogen Narcosis?' Thought Co: Lifelong Learning. Online. Available at: www.thoughtco.com/what-is-nitrogen-narcosis-2963052. Accessed 29 August 2017.

Grant, S. (2014) 'Monsters and Myths: The Manifestations of Man's Deepest Fears'. In: A. Farquharson and M. Clark (eds) *Aquatopia: The Imaginary of the Deep*. London: Tate Publishing, pp. 24–33.

Hamilton-Paterson, J. (2011) *Seven Tenths: The Sea and Its Thresholds*, 2nd Edition. London: Faber and Faber.

Helmreich, S. (2009) *Alien Ocean: Anthropological Voyages in Microbial Seas*. Berkeley: University of California Press.

Helmreich, S. (2011) 'From Spaceship Earth to Google Ocean: Planetary Icons, Indexes, and Infrastructures', *Social Research*, 78(4): 1211–1242.

Hoare, P. (2014) *The Sea Inside*. London: Fourth Estate.

Jameson, F. (1994) 'Contradictory Spaces: Pleasure and the Seduction of the Cyborg Discourse', *Electronic Journal of Virtual Culture*. Online. Available at: http://serials.infomotions.com/aejvc/aejvc-v2n01-jamison-contradictory.txt. Accessed 2 December 2017.

Kurra, S., Lahtinen, K. and Nissinen, A. (2015). *Freediving*. N.p.: Deep Ideas.

Lambert, D., Martins, L. and Ogborn, M. (2006) 'Currents, Visions and Voyages: Historical Geographies of the Sea', *Journal of Historical Geography*, 32(3): 479–493.

Latour, B. (2004) 'How to Talk about the Body? The Normative Dimension of Science Studies', *Body and Society*, 10(2–3): 205–229.

Löfdahl, P., Andersson, D. and Bennett, M. (2013) 'Nitrogen Narcosis and the Emotional Processing during Compressed Air Breathing', *Aviation, Space and Environmental Medicine*, 84(1): 17–22.

Mack, J. (2011) *The Sea: A Cultural History*. London: Reaktion Books.

MacLarnon, A. and Hewitt, G. (1999) 'The Evolution of Human Speech: The Role of Enhanced Breathing Control', *American Journal of Physical Anthropology*, 109(3): 341–363.

MacNoughton, J. and Carol, H. (2016) 'Breathing and Breathlessness in Clinic and Culture: Using Critical Medical Humanities to Bridge an Epistemic Gap'. In: A. Whitehead and A. Woods (eds) *The Edinburgh Companion to the Critical Medical Humanities*. Online. Available at: www.ncbi.nlm.nih.gov/books/NBK379257/. Accessed 14 January 2018.

Margolin, V. (1995) 'The Politics of the Artificial', *Leonardo*, 28(5): 349–356.

McLaren, N. (2017) 'The Martini Effect in Scuba Diving'. Thought Co: Lifelong Learning. Online. Available at: www.thoughtco.com/martini-effect-in-scuba-diving-2963261. Accessed 10 November 2017.

Mentz, S. (2009) 'Toward a Blue Cultural Studies: The Sea, Maritime Culture, and Early Modern English Literature', *Literature Compass*, 6(5): 997–1013.

Merchant, S. (2011) 'Negotiating Underwater Space: The Sensorium, the Body and the Practice of Scuba-Diving', *Tourist Studies*, 11(3): 216–234.

Morgan, E. (1989) *The Aquatic Ape: A Theory of Human Evolution*. London: The Guerney Press.

Néry, G. and Gautier, J. (directors) (2010) *Free Fall*. BlueNery Productions. Online. Available at: www.youtube.com/watch?v=uQITWbAaDx0. Accessed 14 December 2017.

Pell, S.J. (2012) 'Every Space Project Needs an Artist', *Artlink*, 32(3): 12–17.

Pell, S.J. (2014) 'Aquabatics: A Post-turbulent Performance in Water', *Performance Research*, 19(5): 98–108.

Pell, S.J. and Mueller, F. (2013) 'Gravity Well: Underwater Play'. Online. Available at: www.researchgate.net/publication/250308263_Gravity_well_underwater_play. Accessed 4 April 2018.

Peters, K. (2010) 'Future Promises for Contemporary Social and Cultural Geographies of the Sea', *Geography Compass*, 4(9): 1260–1272.

Picken, F. and Ferguson, T. (2014) 'Diving with Donna Haraway and the Promise of a Blue Planet', *Environment and Planning D: Society and Space*, 32(2): 329–341.

Pickering, A. (1999) 'Japanese Eels and Global Warming: A Posthumanist Perspective on Society and the Environment', *Ethics and the Environment*, 10(2): 29–43.

Pine, B.J. and Gilmore, J.H. (1998) 'Welcome to the Experience Economy', *Harvard Business Review*, 76: 97–105.

Probyn, E. (2011) 'Swimming with Tuna: Human–Ocean Entanglements', *Australian Humanities Review*, 51. Online. Available at: http://australianhumanitiesreview.org/2011/11/01/swimming-with-tuna-human-ocean-entanglements/. Accessed 4 April 2018.

Probyn, E. (2015) 'Listening to Fish: More-than-Human Politics of Food'. In: P. Vannini (ed.) *Non-representational Methodologies: Re-envisioning Research*. New York: Routledge, pp. 72–88.

Rozwadowski, H. (2005) *Fathoming the Ocean: The Discovery and Exploration of the Deep Sea*. Cambridge, MA: Harvard University Press.

Rozwadowski, H. (2010) 'Ocean's Depths', *Environmental History*, 15: 520–525.

Sachs, W. (1999) *Planet Dialectics: Explorations in Environment and Development*. London: Zed Books.

Serres, M. (1995) [1990] *The Natural Contract*. Trans. by E. MacArthur and W. Paulson. Michigan: University of Michigan Press.

Steinberg, P. (1999a) 'Lines of Division Lines of Connection: Stewardship in the World Ocean', *Geographical Review*, 98(2): 254–264.

Steinberg, P. (1999b) 'Navigating to Multiple Horizons: Toward a Geography of Ocean-Space', *Professional Geographer*, 51(3): 366–375.

Steinberg, P. (1999c) 'The Maritime Mystique: Sustainable Development, Capital Mobility, and Nostalgia in the World Ocean', *Environment and Planning D: Society and Space*, 17(4): 403–426.

Steinberg, P. (2001) *The Social Construction of the Ocean*. Cambridge: Cambridge University Press.

Steinberg, P. (2013) 'Of Other Seas: Metaphors and Materialities in Maritime Regions', *Atlantic Studies*, 10(2): 156–169.

Szerszynski, B. (2016). 'Planetary Mobilities: Movement, Memory and Emergence in the Body of the Earth', *Mobilities*, 11(4): 614–628.

Vannini, P. and Taggart, J. (2014) 'The Day We Drove the Ice Road (and Lived to Tell the Tale about It) of Deltas, Ice Roads, Waterscapes and Other Meshworks'. In: J. Anderson and K. Peters (eds) *Waterworlds: Human Geographies of the Ocean*. Farnham: Ashgate, pp. 89–102.

Verhaegen, M.J.B. (1985) 'The Aquatic Ape Theory: Evidence and a Possible Scenario', *Medical Hypotheses*, 16(1): 17–32.

Vink, M. (2007). 'Indian Ocean Studies and the "New Thalassology"', *Journal of Global History*, 2(1): 41–62.

Wigen, K. (2006) 'Introduction: Oceans of History', *American Historical Review*, 111(3): 717–721.

10 Women mountaineers and affect

Fear, play and the unknown

Jenny Hall

Introduction

In this chapter I set out to explore how women encounter mountaineering and the affective dimensions of the unknown in what is an essentially masculinized environment (Pomfret and Doran 2015; Frohlick 2006; Ortner 1999). Mountaineers pursue risk, seeking adventures that present challenges beyond what is known both physically and intellectually. This unknown becomes a 'frontier' where fear is embraced as part of a process of accessing powerful sensations of achievement and well-being. Importantly, sensations of fear are not perceived in a negative way; indeed, for mountaineers, fear is healthy, acting as a safety valve, and the single most important factor in survival (Brymer and Monasterio 2015: 206). In what follows, I explore how risk (and emotional responses to risk expressed through fear) becomes a form of play. I suggest that risk is a frontier of play, where mountaineers playfully pursue unknown places, using fear as a fail-safe.

To make my case, I first explore how female mountaineers approach the emotion of fear, drawing on *in-situ* ethnographic accounts and using Derridean notions of 'difference', 'play' and 'presence' as my analytical framework. Second, I consider if female mountaineering is subject to the affective forces of a masculinized culture and what that might mean in terms of their mountaineering practices. I write as a mountaineer and acknowledge that my encounters with other women mountaineers are expressed as co-produced *in-situ* experiences. Through experimental mobile video ethnography, I explore how women mountaineers create a place of difference by negotiating risk when engaged in practices of the unknown. Ultimately, the aim of this chapter is to problematize the seemingly dominant masculine culture of mountaineering, positing that expressions of female difference may constitute legitimate 'other' forms of mountaineering. I also seek to reveal new sentient and emotional geographies of women who mountaineer, with implications for opening both a social and political voice.

Unknown, a place of play

Mountaineering has long been acknowledged as a masculine endeavour. Historically, it provided white Western men with an opportunity to 'perform adventurous masculinities' (Beedie 2015; Frohlick 2006: 109), with many accounts of the heroic deeds performed by men dominating bookshelves, cinemas and other media. What is evident is that women have suffered structural oppression within mountaineering culture. Ortner (1999) describes mountaineering as a form of hyper-masculinization, contending that it has been a predominantly male sport, built on male styles of interaction and all-male institutions. As Pomfret and Doran (2015: 150) state, 'mountaineering landscapes have evolved as, and remain, strongly masculinized concepts … mountaineering has been, and still is, male dominated, despite more women participating in this adventure activity'. Little, it would seem, has changed since Sharp's (2001) research identified that mountain leadership programmes are imbued with a culture of competition, confidence, strength, toughness and proving worth. Masculine leadership traits continue to shape perceptions of who can succeed in a mountaineering space (Barratt 2010; Robinson 2008), whereas women leaders exhibit qualities of cooperation, emotion, personal development, consensual decision-making and being relationship orientated. Sharp (2001) contends that these differences contribute to a perception that women cannot 'cope' or have insufficient physical strength to endure the challenges of 'hard' mountaineering. Yet, the extraordinary achievements of world-class mountaineers like Alison Hargreaves and Julie Tullis, both of whom ascended many high-altitude peaks in the 1980s, contradict this viewpoint (Pomfret and Doran 2015; Beames and Pike 2008).

Research that explores how women experience extreme sports in leisure studies and affectual geographies is limited, particularly in terms of how affect influences sentient and emotional behaviours (Humberstone 2016; Warren 2016; Waitt and Clifton 2013; Evers 2004, 2006, 2009). As Humberstone (2011: 96) asserts, 'the body's absence is echoed by sociologists, albeit frequently gender-blind, exploring physical activity and bodily performance'. Further, within harder forms of mountaineering, key gaps exist concerning the role that gender plays in mountaineering tourism (Pomfret and Doran 2015). This is particularly acute, I argue, with regard to explorations of how women negotiate sensoria and emotion in mountainous regions (see Figure 10.1). In the following sections, I consider how gendered experiences of mountaineering, from a female perspective, might offer insights into different ways of doing mountaineering. I do this in four key ways: first, I contextualize how risk is foundational to mountaineering through Lyng's (2004) concept of edgework; second, I consider Derridean theories of presence, play and difference; third, using two case-studies from my research, I explore the affective dimensions of fear as a practice of play; and finally, I consider how women use fear as a tactical emotion to negotiate positionality within mountaineering. Exploring affect within mountaineering offers a means of building knowledge about how women utilize resistance or coping strategies that may signal different ways in which women do mountaineering. Thus, this becomes revealing of gendered experiences of risk and what this might mean for those who participate in extreme forms of tourism.

Figure 10.1 Alpine day in Scotland
Photograph by Louis Hume

Edgeworking: acting hard

Participation in mountaineering is a radical form of escape from everyday routines, pursuing 'otherness' through sensations of risk; or, as Lyng (2004) describes it, 'edgework'. Lyng's (2004) conceptualization of 'edgework' points to the sensations associated with life-and-death situations and how these act to disrupt the social self (the 'me'). He describes the affect of fear and the resultant 'instinctive' reactions to life-threatening moments, concluding:

> under completely novel conditions of the true edgework, imaginative rehearsal ceases, the 'voice of society' is silenced and the 'me' is annihilated. What is left in place of these elements is a residual 'acting' self that responds without reflective consciousness.
>
> (Lyng 2004: 362)

Thus, risk-takers describe their experiences of pushing the boundaries of fear as self-actualizing or self-determining, authentically real and creatively satisfying. Participation in mountaineering is, arguably, a radical form of play.

How edgeworking is converted into *being*, self-actualizing or self-determining begs the question as to whether the practice of edgework itself is a means of creating symbolic capital. The notion of the residual 'acting' self, I argue, is foregrounded by the affective masculinized impulses to prove worth. Scholars have also critiqued edgework in terms of gender because it focuses on the experiences of white, middle-class, adult males, and therefore does not necessarily account for women's differential perceptions and negotiations of edgework (see Olstead 2011: 86 for a fuller description). Further, it does not allow for how women negotiate the social 'norms' associated with edgework; or, in other words, how women navigate the socio-cultural expectations that they should be risk-averse and not seek opportunities to take risks. While Olstead's (2011) analysis concludes that women's negotiations of risk are only partially successful, she does point out that women who actively seek out risk or edgework do so as a political act that challenges gendered perceptions that women can be part of a culture that takes risks and plays with unknowns. This opens up the possibility for women's participation in extreme mountaineering to be a practice of resistance that enables women to achieve a sense of fulfilment (Olstead 2011).

The women considered in this chapter are described as 'hard' professional mountaineers (Pomfret 2006). Such mountaineers have a tendency to be individualistic, embracing risk, uncertainty and a purity of engagement to achieve a personal aesthetic experience, as opposed to those who mitigate the sensations of risk by being guided by more experienced others (Boyes and Houge Mackenzie 2015). The motivational and therefore behavioural characteristics of 'hard' mountaineers belie two key tensions: first, as Manning (2011) identifies, mountaineers want to escape to the frontiers of experience to achieve a sense of the unknown, but they are continually challenged by their socio-cultural responsibilities towards 'others'; and, second, the perception that mountaineering practices 'other' than a male approach are viewed negatively (Wegner *et al.* 2015; Frohlick 2005; Beedie 2003; Ortner 1999). These tensions hinder engagement in 'other' forms of mountaineering. Normative practices about how to *do* mountaineering bound by 'masculinized' perceptions of a *hard* approach are therefore positional, differentiating those who engage in 'hard' mountaineering from others by generating masculinized symbolic capital associated with experiencing risk (Beedie 2003). How such femininities and masculinities could be expressed are considered through Derridean notions of play, presence and indeterminacy in the following section.

Feminine indeterminacy: different ways to play

> Your eyes are wide open but you cannot see what is coming next, you feel and hear the pressure building and your being reverberates with the immense noise, it hits you and you lean to take the force, standing your ground and pressing hard against an invisible wall, one false step and it will take you over, spindrift lashes at exposed flesh and rucksack straps whip into your face. Snow is moving, picked up and deposited in wave-like formations, sastrugi sculptures, the visible carvings of the wind. Face burning and pinching with cold, eyes

watering, your breath is sucked away, making speaking impossible against the noise and lack of oxygen.[1]

Mountaineers seemingly push the boundaries of 'being' alive by setting goals and taking risks that extend the limits of personal – and, in the case of pioneers, collective – achievement. The visceral stories, where the body is immersed in an act of survival gripped by physical reactions to risk and fear, are formed of the unknown, constituted through sensations that are steeped in uncertainty, and in full presence (Derrida 2006; see Chapters 5 and 8, this volume; see also Figure 10.2). For Derrida, 'play is the disruption of presence' (Derrida 2006: 369); building on this, I pose the unknown as an element that signifies a system of differences and the movement of a chain in which female mountaineering is formed through the practices of play. As Derrida states, 'Play is always play of absence and presence' (Derrida 2006: 369), where we seek to be absent or to escape from everyday life. Absence is where dreams of presence are formed and where the impossible is crafted into a live mountaineering project. In dreaming of presence, as Derrida (2006: 369) asserts:

> Being must be conceived as presence or absence on the basis of the possibility of play ... For there is a sure play: that which is limited to the substitution of given and existing, present pieces. In absolute chance,

Figure 10.2 Mixed winter climbing
Photograph by Louis Hume

> affirmation also surrenders itself to genetic indetermination, to the seminal adventure of the trace.

Those traces are formed of our sentient and emotional experience (Anderson 2009) and through the dreams of a mountaineer's next adventure. Drawing from Derrida (2006), Rose (2006) describes 'dreams of presence' as marking both an imagination of, and a movement towards, presence and a means of attempting to hold on to a world that always eludes our grasp. Play is where presence is felt and, as Rose (2006: 545) asserts, the 'affective cabling that connects self and world'. Rose's (2006) conceptual development of Derrida's (2006) dreaming of full presence synonymizes emotion with affect; the cognitive with bodily sensation; the physical with the psychological; and the cultural with the biological apparatus, to act as one where the processes of affect are not separate or externalized but experienced from within.

While a mountaineer's perception of their world is shaped by many people, places, objects and ideas, their sense of a mountain-place can be known only to them. They can share fragments with others, but these fragments are subject to different interpretations. As Grosz (2005: 91) argues, 'difference is at the very heart of materiality, of ideality … as it is the core of being, and its differentiation in becoming'. One such aspect of difference is gender. Grosz's (2005) analysis of Derrida's work on difference draws out a feminist perspective that opens a pathway to transcend the politics of equality. Derrida rejects a notion of equality between men and women in what he describes as a 'reactive' feminism that could not be resolved through the acquisition of certain rights or values in a bid to make women and men the same. Avoiding binaries and dualisms of this nature, he works on the principle of a fundamental indeterminate difference between the sexes. Through interrogating the very condition of sexuality, he works from the position that women have been attributed a secondary social status on the basis of biological, natural or essential qualities. Derrida understands difference to be the methodology of life, determinant of specificity, and should not be reduced to binaries such as equality, and in doing so he offers a model built on indeterminacy. Thus, indeterminacy frees gender from the considerations of difference before the imposition of sexual difference. Hence, it cannot be captured by sexual identity and exerts resistance to a binary model (Grosz 1995: 121).

Applying indeterminacy to the ways in which women do mountaineering, and specifically how women play at the boundaries of risk, allows detachment from a politics of equality, offering a pre-socialized possibility to consider how women's mountaineering is characterized materially (Grosz 1995). Therefore, Grosz's approach to Derridean theories of difference provides a theoretical foundation to interrogate the affective cabling between the corporeal and perceptual experiences of women, and how difference is manifest. The residual 'acting' self finds a place in indeterminacy or the unknown. Indeterminacy is a space where women can play with risk, if only fleetingly. To experience the unknown is not a silent practice; experiences of risk are latterly shared with peers in a form of trade. Through empirical research, I suggest in the following section that these experiences of the unknown are shaped by the affective dimensions of the masculine culture of mountaineering.

Researching the social life of women mountaineers

One of the greatest challenges when researching the affectivity of social life is how to identify phenomena 'that suggest the presence of affectivity in the material' (Knudsen and Stage 2015: 3). One approach that makes research questions much more answerable is if they are linked to specific bodies in specific empirically approachable social contexts (Wetherell 2015; Knudsen and Stage 2015; Ahmed 2004). In contrast to scholars such as Thrift (2008), who have considered affect as an outside stimulation that first hits the body and then reaches the cognitive apparatus, and is thus not detectable through language, I employ an embodied practice-based model. Methodologically, I draw upon the work of Sara Ahmed (2004), who posits that emotions are synonymous with affect. Ahmed's approach to the cultural politics of emotion and affective economies provides a foundation for considering the connections within a community and how some bodies become marginalized. With her work in mind, I aim to ground embodied experiences of female mountaineers within the complex, live and often problematic personal and interpersonal practices of mountaineering (after Wetherell 2015). I draw upon the work of Ahmed and Wetherell to garner insights from both cultural and psychological perspectives. This mixed approach enables consideration of 'the affective states and meaning-making processes involved in the transmission of affect' (Wetherell 2015: 139; see also Chapter 2, this volume) that female mountaineers experience.

Following the work of Watson and Waterton (2015), Spinney (2015) and Brown (2008), all of whom use methods-in-motion to identify the 'invisible grammars of signification of human behaviour' (Knudsen and Stage 2015: 11), I employed embodied fieldwork to collect *in-situ* and *post-situ* data, from both the participants' and my own perspective, using experimental mobile video ethnography that produced in-depth ethnographic case-studies. The participants and I wore Go-Pro cameras to capture our mountaineering experiences in tandem. The aim was to collect non-semiotic forms of response, or the sensory as well as the semiotic, to understand how women experience a hyper-masculinized culture of mountaineering and determine how difference and voice are negotiated in extreme tourism settings. I asked eight participants to join me in a day of mountaineering, followed by a semi-structured interview that explored key events and topics of conversation from the day. The joint days were co-produced experiences (Oakley 2005) in which I took an active role, and they were varied to accommodate the dynamic nature of weather, which ranged from full days out climbing routes in winter conditions to walking over mountains in 70 m.p.h. winds. My experiences were also supplemented by reflexive field notes.

While the pursuit of mountaineering is dominated by Westernized, white, middle-class, neoliberal, able-bodied men, these are not the sole participants, though they form a 'cultural core' of the practice and shape the social 'norms' of the community (Fletcher 2014: 3). Most of the women who participate in mountaineering, including myself, are similarly Westernized and middle class, and the research participants all shared this background. They were aged between twenty-five and seventy-five, were based in the UK and worked in the mountaineering tourism sector as guides. Of course, not all women mountaineers are white, middle-class Westerners, but my personal experience as an amateur mountaineer facilitated access to this particular social and working environment as an 'insider'. I developed a reflexive approach, recognizing that both I and the research are shaped by my prior and extensive knowledge of mountaineering, and subject to the cultural conditioning and identity of this community. Following Oakley (2005), I sought a methodological approach that would provide the most reliable and democratic way of bridging the gap between researcher and participant: inflict as little harm and provide as much benefit as possible to the participants, but more importantly enable them to reveal their experiences.

Feminine literature describing adventures into the sensory and emotional is rare. One of the earliest accounts was produced by the mountaineer Henriette D'Angeville. Following her ascent of Mont Blanc in 1838, she urged women mountaineers to document their achievements, arguing that it was important for them to write the 'feminine stamp' or feminine experience of mountaineering (Stockham 2012: 2). The literature written by women mountaineers has yet to reach a 'feminine stamp', with one of the few examples being Shepherd's (2011) description of her life-long passion for the Cairngorms. Referring to her sensorial and emotional experiences as the *totalizing* affect of the mountain, she states:

> under me the central core of fire from which was thrust this grumbling grinding mass of plutonic rock, over me blue air, and between the fire of the rock and the fire of the sun, scree, soil, water, moss, flower and tree, insect, bird and beast, wind, rain and snow – the total mountain. Slowly I have found my way in.
>
> (Shepherd 2011: xxv)

Shepherd describes an instinctive, kinaesthetic connection with a specific mountainous area of the Scottish Highlands, where her embodied experience is indeterminate and shaped by chance. In pursuing absence from the everyday, she describes a memory, a 'dream of full presence', where she feels part of a place and 'becomes' a part of the assemblage of a *total mountain*. Humberstone (2013: 566) refers to this as a 'mindfulness of the body when practicing adventurous activity'. Shepherd (2011) resists the urge to conquer summits, believing this misses the point of mountaineering. She expresses a necessity to feel immersed, co-joined and an integral part of a mountain through wandering and encountering the details of rock, flower, insect and weather. This resistance to a more masculine approach to mountaineering, I argue, is an example of how a woman contests what mountaineering could be by impressing on us a different way to experience the mountains. Shepherd embraces her own fragility, exposing her senses to her feelings of insignificance and vulnerability, a small cog in the vast wheel of the Cairngorm plateau. By controlling her fear, she achieves a profoundly sentient experience of being part of the *totalizing* mountain. Her vulnerability is a space of play, which I suggest reveals a potentially different kind of feminine mountaineering practice.

Of course, this is not an exclusively female phenomenon, but part of a deeply ingrained history born of the desire to experience a sense of awe, otherness and terror, or the sublime. This was popularized through the philosophical and artistic impressions of the late-eighteenth- and early-nineteenth-century Romantics, such as Wordsworth and Coleridge, who drew inspiration from excursions the English Lake District (Picard 2016). Beedie (2015: 50) discusses how the 'Romantic movement became responsible for creating and then sustaining ideas about discovery, exploration and the search for an authentic self.' However, a dominant, contemporary, Westernized ideology of 'conquest' masks other forms of mountaineering experiences, such as seeking the sublime. Post-war nationalism fuelled a global race to be the first to ascend the world's highest peaks, which has impacted significantly on public perceptions of what mountaineering should be, and how to experience mountains and mountaineering as *tourists.* This led to the development of mountaineering as a form of masculinized symbolic capital, born of a desire to acquire experiences and achieve adventurous challenges established by pioneers of routes previously believed to be impossible. Everest is a prime example of how the most inhospitable and extreme challenges have become commodities in a booming adventure tourism industry (Beedie 2015).

Successful mountaineers secure lasting, useful relationships that can generate symbolic, social and material profits (Bourdieu 1989). The symbolic profits a mountaineer seeks to acquire are positional, securing a 'name' as a 'hard' mountaineer. This name positions the mountaineer to reap social and material capital, ensuring financial survival, access to resources (such as people, equipment, contacts and geographical locations), exclusive groups and clubs, and opportunities to pursue new projects. By way of explanation, the social relationships produced individually, collectively, consciously or unconsciously establish and reproduce the symbolic capital that secures a place within a mountaineering community (Bourdieu 1989). Beedie (2015) considers the desire to escape and achieve a sense of immersion or otherness from everyday urban life in contrast to the desire to extend status, compete and accumulate symbolic capital through experiences as an ongoing social 'battleground'. The tensions arising from this battleground are evident in the social life of female mountaineering, and I turn to them now. Through two case-studies, I consider how playing with risk is underpinned by the acquisition of masculinized symbolic capital.

Case-study One

For the eight participants involved in my research, happiness was rooted in pushing themselves to accept challenges. The word 'happiness' itself was rarely used; instead, terms such as 'well-being', 'achievement', 'confidence', 'self-reliance' and 'self-esteem' were cited as reasons for pushing beyond the known. When discussing the risks involved, the participants talked about the transient nature of achieving sensations of 'in the now' and 'flow'. For instance, Lorrie[2] described a day when she went out alone to climb 'solo' – without any of the protection afforded by ropes and other climbing equipment – and noted how the sensations of movement across the rock transcended any sense of fear:

> I have had some incredible experiences where it's just you and the activity, and feel a sense of flow, a sense of being at one. You don't worry ... fear doesn't come into it, as long as I am in the mood, the movement just flows ... It's that thing of being totally in the now ... and somehow feels like the right thing, where your mind and body are all kind of in the same place, wanting the same thing for you.
>
> (Lorrie, interviewed 18 November 2016)

Here, Lorrie is describing a sensorium that incorporates a much wider array of stimuli and responses than the five most recognized senses: hearing, smell, taste, touch and vision. Sather-Wagstaff (2017) describes this as a polysensorial experience that incorporates the kinaesthetic, from proprioception – the biological sensation of a body in space – combined with the culture-specific biography of previous experiences and interpretation. As Sather-Wagstaff (2017) points out, that which fails to be explained is the impact of affective forces arising from everyday life. There are many questions arising from Lorrie's solo climbing activities, but I want to focus on how fear acts as a frontier for survival when exposed to extreme risk, and how Lorrie's climbing required a polysensorial space in which to play that was free from feelings of being frightened. Moreover, it needed the indeterminacy of the unknown, an absence generating a curiosity for pushing a frontier of polysensorial knowledge connecting human and non-human through movement.

The excitement with which Lorrie's stories of physical endeavour were related were personally thrilling, generating an infectious desire to pursue my own mountaineering projects. The affective dimensions of the sublime, or the need for feeling a connected otherness and awe, oscillate at the brink of the impossible or the unknown, where fear is a playful tool for pushing beyond a muscular and emotional frontier. For one fleeting moment, Lorrie was free from the everyday; she had carved a space to be a mountaineer. Yet, I contend that her experience was motivated by a need to *conform* to a culture of mountaineering by appearing fearless, something that could be traced by the animated way in which the story was told, creating symbolic capital in terms of being 'hard' and proving her worth.

Lorrie also talked about experiencing a wonderful sense of in the now on the first day of a climbing trip, then losing the sensation the following day:

> I had a very good feeling one particular day a few winters ago, where I went and soloed a number of ice routes and I started slowly climbing harder and harder routes and feeling amazing. Then went back a second day and was going to do the same, but started feeling, 'I don't think I should do this.' I did one route and then walked back out. I felt like it had gone past the point of being about me and was becoming … an addictive kind of thing that felt like I might be pushing my limits here.
>
> (Lorrie, interviewed 18 November 2016)

Lorrie attributed 'changing her mind' – and losing the sensation of being in the now – to sharing her experiences with her mountaineering peers on the first evening. Competition crept in, with her fellow-climbers commenting on what she should climb and how she should do it. This ruptured Lorrie's sensation of in the now and the expedition became 'about something else' – about proving her worth. Lorrie's account contains an undertone of social conformity and points to the question of how women negotiate space that allows the body to achieve sensations of in the now. In her case, she found space by soloing. Thus, reflexive assessment takes over when a body experiences emotion through

Figure 10.3 Topping out
Photograph by Louis Hume

storytelling, and normative behaviours curb the risks a mountaineer is prepared to take. I posit that masculine cultures of competitive storytelling foreground the need to seek sensations of in the now. In other words, sensations of in the now are acheived through playing with risky unknowns that lead to the creation of a specific form of feminine symbolic capital (see Figure 10.3).

Case-study Two

The weather was horrendous throughout the day I spent with Freddie (7 December 2016), with one forecast predicting significant precipitation and gusts of up to 96 m.p.h. on the summits of the Cairngorms. Freddie said she would consult at least three different forecasts to get a sense of what the weather was doing and then make a note of which was the most accurate. This gave her a sense of confidence about which one to use for subsequent days out on the hills. We chose to do a lower-level route, going up to a height of 732 metres – a relatively easy day in terms of both distance and climbing – because the strong wind meant conditions were too bad to attempt anything else. There was no hesitation or perceivable worry on Freddie's part about venturing into what many would consider an extremely hazardous environment.

Halfway through our climb, we sheltered behind a rock to discuss the route and how the wind would affect us. Freddie pointed out that we would be in the lee on the eastern side of the hill until we met the full force of the wind at the second col, but thereafter it would push us up to the final peak. By that stage, the wind had already caught Freddie's hood and blown her over, demonstrating the severity of the conditions. She offered two alternatives for our descent: either head back towards the forest (the easier route) or battle back over the ridge and into the wind, which was now strengthening. We had to shout to make ourselves heard:

JENNY: I don't really feel the need to do that [walk into a 70 m.p.h. wind].
FREDDIE: It would be fresh [laughing with a goading intonation].
JENNY: Yes, a bit blustery [not laughing and trying to disguise my alarm]
FREDDIE: Yeah, a blustery day [laughing hard].

Understating the severity of a situation is ubiquitous in mountaineering language. It is a code of behaviour based on the belief that mountaineers can take 'whatever the mountain throws at us' (Lorrie, interviewed 18 November 2016). If I hadn't known better, this hard-faced attitude might have created an atmosphere in which I would have agreed to head back over the ridge just to prove I was capable. It reveals an underlying power dynamic that can, and often does, lead to what McCammon (2004) identifies as 'heuristic traps' or decision-making processes based on 'rules of thumb' rather than hard facts. McCammon (2004) identifies heuristic traps of familiarity, scarcity, social proof and commitment as key factors in human-triggered avalanches, and his findings are widely understood and shared within the UK's national mountaineering centres through their training programmes. The heuristic thought processes in mountaineering are dominated by a masculine culture of competition, 'hardcore', fearless and 'getting away with it' (Frohlick 2005), and I experienced this on the mountain with Freddie. Indeed, it was difficult to overcome the competitive urge to prove myself (McCammon 2004) and show I was sufficiently 'hard' in that particular social situation.

Until that point, Freddie and I had enjoyed playing in the wind, 'whooping' and 'whoahing' as it demonstrated its power, excited at 'getting away with it' and being in the now, but the moment was broken by my fearfulness and resistance to being goaded. I conjured up memories of being blown over that grew into visions of being dashed on the rocks or forced over a precipice, my fear acting as a fail-safe that told me it was time to turn back. A keen sense of inferiority crept in as I felt my inability to meet the challenge. I had been put in a position to choose the lesser option and I had demonstrated weakness by doing so. I was not up for pushing any further into the unknown, and in that moment Freddie had asserted her dominance by forcing me to make the decision. At the beginning of the day, she had exhibited a feminine leadership style by carefully researching and collaboratively planning the journey with me and gauging all of the environmental and human factors (Sharp 2001). But now she had moved into a heuristic model of leadership – what Dilley and Scraton (2010) describe as 'female masculinity' – by literally throwing caution to the wind. I suggest that this is driven by a continual need to prove 'you are good enough' (Jo, interviewed 8 December 2016) in a masculinized culture.

The masculinized culture of mountaineering requires women to work extremely hard both physically and socially to achieve and maintain their status as competent leaders, which makes them very protective of their 'names'. As Jo (interviewed 8 December 2016) notes, 'Having a name is actually quite a thing in the outdoor industry.' This pressure to demonstrate high performance in extreme environments was palpable even amongst those participants who are considered to be at the height of their profession, having summited 8,000-metre peaks and pioneered new routes. Attitudes towards this competitive atmosphere were expressed as a means to 'make the grade' (Lorrie, interviewed 18 November 2016) – to be accepted but not dominant. The challenges they set themselves were introverted personal goals rather than overt public expressions of achievement. Annie (interviewed 6 December 2016) admitted that she would 'go at it like a bull in a china shop in a bid to prove I could do it as well as the men' in her younger days.

For these women, the processes of playing with risk even when seeking in the now experiences is underpinned by a desire to prove their worth in a highly competitive mountaineering culture in a bid to feel accepted, demonstrate their strength and competence, and achieve a 'name'. The affective traces of masculinized culture are evident in their attitudes towards competition and status, and in Freddie's switch from a collaborative style of leadership to a female form of masculinized leadership, which signalled her desire to acquire symbolic capital and reinforce her 'name'. The affective dimensions of a woman's mountaineering experiences in both professional and personal contexts are thus bound in a masculinized form of risky play. As Sather-Wagstaff (2017: 17) suggests, being in the now is the 'essence of the sensorium as a highly complex, culturally mediated, and thus varied, biologically ground (but not biologically determined) processual meaning-making phenomenon'. According to Lyng (2004), women's sentient experiences of mountaineering seem to annihilate the 'voice of society' and the 'me', but this does not account for how women negotiate – or why they choose to pursue – their mountaineering experiences, or why they push so hard into the unknown. I contend that every level of a woman's mountaineering is contaminated (Knudsen and Stage 2012) by the masculinized atmospheric flows of how to do mountaineering, and I posit that the phenomenon of being in the now is the product of acquiring the symbolic capital associated with proving a 'name'.

Conclusion

Fear keeps us alive; it is our fail-safe and it is also our frontier for self-discovery leading to feelings of self-esteem and achievement. But is not an escape or a place of freedom. The fleeting nature of being in the now of the 'residual acting self' (Lyng 2004) is bound by atmospheres that percolate through the socio-cultural dimensions of the masculine environment of mountaineering. Fear is the medium, but play is where we achieve our dreaming of full presence, where being in the now enables a connectedness to journey into the unknown, formed of rock, air, human and the non-human – the *totalizing mountain*. Yet, the very act of being in the now is a product of the affective dimensions of building symbolic capital that proves a mountaineer's sense of place within a mountaineering culture. Tracing the way women mountaineers negotiate the competitive forces exerted upon their mountaineering encounters reveals a different feminine geography of women who mountaineer, opening the potential for a new voice.

Notes

1 Author's field notes, 17 November 2016.
2 Pseudonyms have been used to protect the privacy of all four of the research participants who are quoted in this chapter: Lorrie, Annie, Jo and Freddie.

References

Ahmed, S. (2004) *The Cultural Politics of Emotion*. New York: Routledge.
Anderson, J. (2009) *Understanding Cultural Geography: Places and Traces*. London: Routledge.

Barratt, P.R. (2010) 'Hybrid Climbing Bodies: The Climbing Assemblage and the Technologically Mediated Engagements and Ascensions of Rock Climbers'. Ph.D. thesis, University of Hull.

Beames, S. and Pike, E.C.J. (2008) 'Goffman Goes Rock Climbing: Using Creative Friction to Explore the Presentation of Self in Outdoor Education', *Australian Journal of Outdoor Education*, 12(2): 3–11.

Beedie, P. (2003) 'Mountain Guiding and Adventure Tourism: Reflections on the Choreography of the Experience', *Leisure Studies*, 22(2): 147–167.

Beedie, P. (2015) 'A History of Mountaineering Tourism'. In: J. Highham, A. Thompson-Carr and G. Musa (eds) *Mountaineering Tourism: Activity, People and Place*. London: Routledge, pp. 40–54.

Bourdieu, P. (1989) 'Social Space and Symbolic Power', *American Sociological Association*, 7(1): 14–25.

Boyes, M. and Houge Mackenzie, S. (2015) 'Concepts of the Wilderness Experience and Adventure Mountaineering Tourism'. In: J. Highham, A. Thompson-Carr and G. Musa (eds) *Mountaineering Tourism: Activity, People and Place*. London: Routledge, pp. 66–84.

Brown, K.M., Dilley, R. and Marshall, K. (2008) 'Using a Head-mounted Video Camera to Understand Social Worlds and Experiences', *Sociological Research Online*, 13(6). Online. Available at: www.socresonline.org.uk/13/6/1.html. Accessed 18 March 2018.

Brymer, E. and Monasterio, E. (2015) 'Mountaineering Personality and Risk'. In: J. Highham, A. Thompson-Carr and G. Musa (eds) *Mountaineering Tourism: Activity, People and Place*. London: Routledge, pp. 198–210.

Derrida, J. (2006) *Writing and Difference*, 7th Edition. Trans. by A. Bass. London: Routledge Classics.

Dilley, R.E. and Scraton, S.J. (2010) 'Women, Climbing and Serious Leisure', *Leisure Studies*, 29(2): 125–141.

Evers, C. (2004) 'Men Who Surf', *Cultural Studies Review*, 10(1): 27–41.

Evers, C. (2006) 'How to Surf', *Sports and Social Issues*, 30(3): 229–243.

Evers, C. (2009) '"The Point": Surfing, Geography and a Sensual Life of Men and Masculinity on the Gold Coast, Australia', *Social and Cultural Geography*, 10(80): 893–909.

Fletcher, R. (2014) *Enhancing the Wild: Cultural Dimension of Ecotourism*. Durham, NC: Duke University Press.

Frohlick, S. (2005) '"That Playfulness of White Masculinity": Mediating Masculinities and Adventure at Mountain Film Festivals', *Tourist Studies*, 5(2): 175–193.

Frohlick, S. (2006) 'Wanting Children and Wanting K2: The Incommensurability of Motherhood and Mountaineering in Britain and North America in the Late Twentieth Century', *Gender, Place and Culture*, 13(5): 477–490.

Grosz, E. (1995) 'Ontology and Equivocation: Derrida's Politics of Sexual Difference', *Diacritics*, 25(2): 114–124.

Grosz, E. (2005) 'Derrida and Feminism: A Remembrance', *Difference: A Journal of Feminist Cultural Studies*, 16(3): 83–94.

Humberstone, B. (2011) 'Embodiment, Social and Environmental Action in Nature-based Sport: Spiritual Places', *Leisure Studies*, 30(4): 495–512.

Humberstone, B. (2013) 'Adventurous Activities, Embodiment and Nature: Spiritual, Sensual and Sustainable? Embodying Environmental Justice', *Motriz, Rio Claro*, 19 (3): 565–571.

Humberstone, B., Fox, K.M. and Dubnewick, M. (2014) 'Cycling into Sensoria: Embodiment, Leisure and Tourism', *Tourism International Review*, 18(1): 71–85.

Humberstone, B., Prince, H. and Henderson, K.A. (eds) (2016) *Routledge International Handbook of Outdoor Studies*. London: Routledge.

Knudsen, B.T. and Stage, C. (2012) 'Contagious Bodies: An Investigation of Affective and Discursive Strategies in Contemporary Online Activism', *Emotion, Space and Society*, 5(3): 148–156.

Knudsen, B.T. and Stage, C. (2015) *Affective Methodologies: Developing Cultural Research Strategies for the Study of Affect*. Basingstoke: Palgrave Macmillan.

Lyng, S. (2004) *Edgework: The Sociology of Risk-taking*. London: Routledge.

Manning, R.E. (2011) 'Motivations and Benefits in Outdoor Recreation'. In: R.E. Manning (ed.) *Studies in Outdoor Recreation: Search and Research for Satisfaction*, 3rd Edition. Corvallis: Oregon State University Press, pp. 166–189.

McCammon, I. (2004) 'Heuristic Traps in Recreational Avalanche Accidents: Evidence and Implications', *Avalanche News*, 68. Online. Available at: www.sunrockice.com/docs/Heuristic%20traps%20IM%202004.pdf. Accessed 18 March 2018.

McDowell, L. (1992) 'Doing Gender: Feminism, Feminists and Research Methods in Human Geography', *Transactions, Institute of British Geographers*, 17: 399–416.

Moffat, G. (2013) *Space below My Feet*. London: Weidenfeld & Nicolson.

Oakley, A. (2005) *Experiments in Knowing: Gender and Method in the Social Sciences*, 2nd Edition. Cambridge: Polity Press.

Olstead, R. (2011) 'Gender, Space and Fear: A Study of Women's Edgework', *Emotion, Space and Society*, 4: 86–94.

Ortner, B.S. (1999) *Life and Death on Mount Everest: Sherpas and Himalayan Mountaineering*. Woodstock: Princeton University Press.

Picard, D. and Robinson, M. (2016) *Emotion in Motion: Tourism, Affect and Transformation*. London: Routledge.

Pomfret, G. (2006) 'Mountaineering Adventure Tourists: A Conceptual Framework for Research', *Tourism Management*, 27: 113–123.

Pomfret, G. and Doran, A. (2015) 'Gender and Mountaineering Tourism'. In: J. Highham, A. Thompson-Carr and G. Musa (eds) *Mountaineering Tourism: Activity, People and Place*. London: Routledge, pp. 138–155.

Robinson, V. (2008) *Everyday Masculinities and Extreme Sport, Male Identity and Rock Climbing*. Oxford: Berg.

Rose, M. (2006) 'Gathering Dreams of Presence: A Project for the Cultural Landscape', *Environment and Planning D: Society and Space*, 24(4): 537–554.

Sather-Wagstaff, J. (2017) 'Making Polysense of the World'. In: D.P. Tolia-Kelly, E. Waterton and S. Watson (eds) *Heritage, Affect and Emotion: Politics, Practices and Infrastructures*. London: Routledge, pp. 12–29.

Sharp, B. (2001) 'Take Me to Your (Male) Leader', *Gender and Education*, 13(1): 75–86.

Shepherd, N. (2011) *The Living Mountain*. Edinburgh: Canongate.

Spinney, J. (2015) 'Close Encounters? Mobile Methods, (Post) Phenomenology and Affect', *Cultural Geographies*, 22(2): 231–246.

Stockham, K. (2012) '"It Went down into the Very Form and Fabric of Myself": Women's Mountaineering Life-writing 1808–1960'. Ph.D. thesis, University of Exeter.

Thrift, N. (2008) *Non-representational Theory: Space, Politics and Affect*. London: Routledge.

Waitt, G. and Clifton, D. (2013) '"Stand out, Not up": Bodyboarders, Gendered Hierarchies and Negotiating the Dynamics of Pride/Shame', *Leisure Studies*, 32(95): 487–506.

Warren, K. (2016) 'Gender in Outdoor Studies'. In: B. Humberstone, H. Prince and K.A. Henderson (eds) *Routledge International Handbook of Outdoor Studies*. London: Routledge, pp. 360–368.

Watson, S. and Waterton, E. (2015) 'Methods in Motion: Affecting Heritage Research'. In: B.T. Knudsen and C. Stage (eds) *Affective Methodologies: Developing Cultural Research Strategies for the Study of Affect*. Basingstoke: Palgrave Macmillan, pp. 97–118.

Wegner, C.E., Ridinger, L.L., Jordan, J.S. and Funk, D.C. (2015) 'Get Serious: Gender Constraints to Long-distance Running', *Journal of Leisure Research*, 47(3): 305–321.

Wetherell, M. (2015) 'Trends in the Turn to Affect: A Social Psychological Critique', *Body and Society*, 21(2): 139–166.

11 'Frontier climbing' in the wild, wild East

Esther Bott

Introduction

> Wilderness hides its unnaturalness behind a mask that is all the more beguiling because it seems so natural. As we gaze into the mirror it holds up for us, we too easily imagine that we behold its nature, when in fact we see the reflection of our own unexamined longings and desires.
>
> (Cronan 1996: 7)

In the above quote, we are invited to consider the paradoxical nature of wilderness, as an entity that is at once essential and subjective, both independent and dependent on human reflection for its existence, in whose seeking we are also searching for ourselves. This chapter explores such tensions in the often contradictory and always subjective notions of 'adventures' and 'frontiers' in overlapping dimensions through the context of rock-climbing tourism in Li Ming in remote southern China. Climbing as a sport can be characterized by the frontier metaphor: as boundaries and limits to be tested and new ground to be broken, which leads climbers into new terrain with regard to difficulty, techniques, technologies and, of course, routes. But the geographical, cultural and economic environment within which those rocks are located are also enmeshed with the quality of the adventure and the measurement of 'wildness' because the remoteness, perceived exoticism and 'difference' of a country or area contribute to the challenges and rewards of adventures. This connects with how the frontiers of tourism are also pushed back, as tourists seek to consume experiences that extend beyond what they perceive as 'normal'. As tourism becomes truly globalized, and a demand for more 'edgy' experiences grows, tourism development is creeping into not only the world's least developed areas, but also into its most geographically peripheral ones. The restlessness of tourists leads to a seemingly endless quest for novel experiences and this increasingly involves cross-cultural interactions with exoticized local people. Interestingly, it is not uncommon for climbers travelling abroad to push the various boundaries of the sport to disavow the category of 'tourist' and actively avoid areas and behaviour that they consider 'touristy' (Bott 2015). Yet, discussions here will nonetheless bring them together at an embodied level (climbers spending time abroad *are* tourists) and at the symbolic and subjective nexus of the adventure/tourism frontier.

China represents a relatively new tourism market for Western visitors. Like many other parts of Asia, its development has been especially rapid in areas where niche touristic practices represent 'the edges of what we would generally take to be mainstream tourism' (Robinson 2005: xix), and within which contemporary tourists are increasingly seeking consumer experiences that transcend the seductions of 'mass' tourism (Cohen 1988). These new, unorthodox or 'niche' areas are being theorized in terms of how the desires and motives underpinning them compare to those related to 'standard' tourism (Lennon and Foley 2002; Novelli 2005; Sharpley and Stone 2009) and how their differentializing qualities might facilitate the processes of 'selfing' and 'othering' related to tourists' quests for novelty (Bott 2015). Adventure tourism, as one important niche, is being conceptualized as an individualized project of 'self', connected to the reproduction of social identities through risk performance (Rhinehart 2003; Palmer 2004; Bott 2009), which can be understood as a form of embodied resistance to the restrictions and limitations imposed by modernity (Lewis 2000). According to Cohen (1988), adventure tourists are 'existential tourists' who reject the normalized, controlled leisure of the masses.

Whilst the numbers of visitors to China for rock-climbing purposes are too low to warrant quantitative attention, or to be considered a significant market in themselves, climbing tourism in Li Ming is revealing in terms of how its minority appeal and the discourse surrounding it feature the notion of the frontier. This connects to wider concerns about the consequences of tourism's expansion into ill-prepared localities. Little is written about Li Ming in the climbing press, but what is, as discussed below, weaves a narrative between the hostile character of the climbing, Li Ming's spectacular, remote scenery, unparalleled scope for adventure, and 'interesting' cultural encounters. This chapter draws on a range of data sources, including climbing articles, guidebooks, blogs and other social media discussions, participants' observations and autoethnography, to explore the nexus of activities, environment and local people in the production of the 'frontier climbing' adventure.

Tourism and identity

Tourism commodifies 'place', meaning that locations and their inhabitants become evaluated, represented and altered for their exchange value. Tourism constitutes consumption, where places hold specific sign and meaning value to individual consumers. Cultural capital is accumulated through travel to specifically valued parts of the world, as practices, experiences and landscapes are aestheticized by 'cultural intermediaries' for the reproduction of distinction (Bourdieu 1984). For example, travelling to remote regions with harsh or hazardous conditions carries a certain amount of cachet. Travelling overland in Uganda, for instance, sends out significantly different messages about the self than taking a two-week package holiday in Mallorca. Similarly, mountains capture the 'romantic gaze' (Urry 2002) in special ways with their untameable, unpredictable qualities, but they are also 'classed' spaces that attract wealthy tourists seeking distinction. Certain prefixes to new forms of tourism – 'eco', 'adventure' and 'independent', for example – carry positive associations for consumers seeking to differentiate themselves from the meanings and problems attributed to mass tourism. Mowforth and Munt (2009: 131) argue that this often occurs as a class-based acknowledgement of tourism as 'an environmentally, socially and culturally problematic activity among certain fractions of the new middle classes'. As a result, they argue, 'the intellectual demands of these class fractions are of increasing importance in the legitimation of travel' (Mowforth and Munt 2009: 131). Distinguishing oneself from mass tourism by travelling further afield to more 'global' and less visited places permits access to globalized spaces that are, in Bauman's (1998: 93) words, geared towards tourists' 'dreams and desires' in increasingly self-conscious ways.

The identity-producing potential of tourism is being explored in a wealth of theoretical and empirical literature. Tourists' decisions about the nature and location of their holidays hold the potential to reproduce positively evaluated, creative, independent identities (Douglas *et al.* 2001; Meethan 2001; Stebbins 2002; Novelli 2005). Further, niche tourism in unusual places can signal a wealth of cultural capital, which often involves, as discussed above in relation to climbers, the disavowal of the 'self-as-tourist'. Further to seeking out new pursuits and practices in order to delimit and distinguish the self, distinction can be emphasized when there is an element of danger attached to the (in)stability of the location, as, for example, in the case of areas prone to natural disasters or political unrest. Timothy (2002) suggests that stepping into places that are perceived as dangerous, or that closely border forbidden, rarely visited or off-limits regions, can boost self-reflexive ideas about adventurousness. Encounters with risk and danger, albeit sometimes more perceived and tokenistic than tangible, have become a new element of what Urry (2002) has termed the 'tourist gaze', by which he means the expectations that tourists place on local populations when they participate in tourism. Thus, essential to the gaze of tourist spaces or experiences that are considered dangerous is the performance of risk.

Conceptualizing adventures

The 'adventure' is a slippery concept that incorporates a vastly differentiated range of subjectively held notions. Adventures are personal and fluid, but also managed and marketized, especially in terms of adventure tourism. Here, perhaps conceptualization becomes trickier yet, because holidays that are themselves not especially risky can occur in 'adventurous' environments. Conversely, organized holidays in a controlled, rationalized environment can involve the possibility of an isolated adventure experience – a beach bungee-jump during a package holiday, for example. Varley's 'adventure commodification continuum' proposes that holidays labelled as 'adventures' range from the 'controlled, rationalized and risk-managed "post-adventure" to the actual or "original" adventure which requires and allows greater levels of commitment, risk and responsibility' (Varley 2006: 173). Landscapes and environments that are considered adventurous are therefore becoming commodified alongside other tourism locations, and this increasingly includes mountain environments. Beedie (2003: 626) argues that mountains are particularly attractive destinations for adventure tourism because they offer 'a range of activity options in a setting steeped in actual and symbolic representations of adventure'. He goes on to argue that:

> [Mountains] are also wild rugged places that attract bad weather and contain objective dangers, such as exposure to extreme elemental conditions and loose rock, which make mountain recreation activities inherently risky and hazardous. However, it is this uncertainty and potential for personal harm that generates excitement by setting such undertakings in a context of challenge and adventure.
>
> (Beedie 2003: 627)

Mortlock (1994) has foregrounded the importance of subjectivity in the formation of adventures and suggests that subjective experience is born of the uniquely configured intersection of the objective physical task (including environmental conditions) and the person's capabilities and resources at the time. The adventure experience, therefore, emerges from the juncture of prior experience and the surrounding circumstances of current adventures. Mortlock identifies distinct states of adventure, which, related to this discussion, include 'frontier adventure'. This state is characterized by what he terms 'peak experience', which emerges from the adventurer meeting challenges very close to their limits. If the person succeeds, then generally a peak experience is achieved, but there is a risk of pushing too far and falling/failing, potentially leading to a state of 'misadventure' (Mortlock 1994). The notion of 'peak experience' is developed in Priest's 'adventure experience paradigm' (AEP). The AEP, like peak experience, is based on the relationship between personal skill level and situational challenge and is also conceptualized as distinct states, one of which, termed 'peak adventure', refers to the point at which personal competence and skill level equal the situational risk; this represents the ultimate goal of self-motivated individuals (Priest 1990). While theoretical critiques of these models include their oversimplification of human desires and motivations and failure to consider social, cultural and economic factors adequately (Nerlich 1997), there is also little investigation into, or emphasis on, the significance of adventurers' direct, embodied relationships with the setting.

The setting: Li Ming and Laojunshan Park

Nestled in remote mountainous terrain in China's Yunnan Province, Li Ming is a recently discovered rock-climbing area in the very early stages of climbing and tourism development (Figure 11.1). The climbing is of a particularly adventurous nature, with long, high routes that are climbed using temporarily placed equipment for protection (as opposed to being permanently equipped with fixed bolts), and rock – cracked sandstone – that is notoriously physical and difficult to climb, as this excerpt from an article in *Climbing* magazine illustrates:

> After a long battle of climbing up and down, learning how to walk cams, and tearing the skin on her leg from knee-barring, she made it to the top. Exhausted, with gritty red sand in her teeth and dirt all over her face, she smiled broadly – you could tell it was part excitement and part relief. It was evident that she was already thinking about what route to get on next. She had tasted adventure, and she was certainly ready for more.
>
> (Dickey 2013)

As climbing – particularly indoor, artificial climbing – grows in unprecedented popularity, the 'extremeness' of rock-climbing and its role as a minority sport are arguably called into question, leading to a potential dilution of its identity-giving qualities. As a result, it can be argued that climbing and outdoor pursuits more broadly have become rationalized and commoditized to such a degree in the West that the heartfelt 'adventure' – as an ephemeral, spontaneous event – is near impossible. This inspires some climbers to seek out new niches in which to specialize in order to remain at the frontier (Bott 2015). Like other types of special-activity tourists, climbers seek out specific 'niche' locations to practise their sport – a certain medium of rock or a newly developed or discovered area, for instance. Climbing itself is also differentiated in many ways, having various sub-categories of techniques and methods, and participation varies in range. Some climbers partake in their sport occasionally, or at weekends and during holidays. Others dedicate their lives to climbing, often forsaking careers, family life, material gain and so on. These 'lifestyle climbers'

Figure 11.1 Li Ming climbing panorama
Photograph by Esther Bott

or 'dirtbags' are 'a subculture of highly dedicated individuals who give up permanent residences for the full-time pursuit of this sport' (Rickly-Boyd 2012: 85). Such climbers tend to value locations in terms of how basic and 'unspoilt' they are; in other words, they appreciate which tourism and infrastructural facilities are *not* offered. Within this particular framework of distinction there is often a desire to travel further afield than the majority, to climb routes that do not attract crowds (of climbers or other tourists) and do not have mainstream climbing press guidebooks, maps and so on. Li Ming attracts precisely this kind of climber. For Westerners, climbing in China and other remote, developing areas in the Global South represents a niche within a niche, offering solutions to the problem of rationalization of adventure.

During our month-long climbing trip to Li Ming in the peak season of 2016, my partner and I encountered only nine other climbers. This meant quietness and solitude for us on an unprecedented scale, in comparison to other climbing areas of similar scope and quality in, for example, Europe or North America. Also, unlike these popular areas, Li Ming still offers many unclimbed opportunities for first-ascensionists to discover, climb, grade and name new lines – a luxury that has been largely consigned to climbing history in the West for all but the elite few who are capable of pushing the upper limits of the sport (Figure 11.2). The sole climbing guidebook for Li Ming (Dobie 2015) is a basic, self-published manual, available only in PDF format

Figure 11.2 Li Ming village
Photograph by Esther Bott

from the author, US climber Mike Dobie, who has been active in the area since 2010 and is credited, alongside a handful of other US and Chinese climbers, as a principal pioneer in its development. The Li Ming climbing website encourages climbers to send Dobie details of new routes and feedback on his route grades and other details, while he provides regular updates in return. This gives the guidebook a distinctly homespun, interactive and democratized feel, which represents a departure from the systematic and rationalized guidebook market in Western contexts. Unsophisticated, small-scale production of guidebooks marks a temporal point in the history of climbing discovery and development, because, generally speaking, the longer an area is established, the more popularity it gains and the more 'professional' and consolidated the literature becomes. The hand-drawn maps, low-resolution photographs and independent format and style of Dobie's guidebook set it apart from the glossy, mass-produced, highly stylized publications that are typical of prominent guidebook publishers, such as Rockfax. The latter's formulaic, expensive volumes generate controversy and a degree of ill-feeling amongst some sections of the climbing community, who express their disapproval of the commoditization of the hard work of climbers and their generous dissemination of knowledge, descriptions, accurate grading and so on. To some degree, Li Ming seems to facilitate a return to a pre-capitalist point in climbing history, when knowledge was shared locally without the presence of profiteering mediators, and details were left vague and sufficiently unreliable to permit authentic adventures.

Disappearing wilderness

The search for 'wilderness' is common in 'frontier' climbing discourse, which can be viewed as seeking to outrun rationalization processes. Wilderness was, of course, the domain of the original 'frontiersmen', epitomized by uncontrolled nature, great physical risk, hardship and solitude. The scope for discovery and self-actualization of the frontier was vast but ultimately led to two inevitable by-products: penetration and destruction. Wilderness, whose main draws are tranquillity and proximity to nature, can quickly and paradoxically lose its appeal precisely because of its appeal. This has very real consequences for inhabitants whose livelihoods might disappear as suddenly as they arose in the fickle market for 'new' tourism, and whose homelands might now be unrecognizable because of tourism development. Swarbrooke (2003) has argued that a similarly paradoxical logic underpins adventure tourism, because, once popularized, the 'adventure' fails to be rarefied and individualistic. In the same way, Simmel's analogous 'love affair' ceases as soon as it begins, because its very existence precludes its mystery; as Simmel knew, 'our general longing for light is to that accidental and immediately disappearing brightness' (Simmel 1965: 248).

Wilderness is now, to a large degree, tamed because the only remaining so-called wilderness is that which has been so designated. In the battle between humankind and nature, often reproduced in literary and artistic renditions of the American frontier and so violently captured in Alejandro Iñárritu's film *The Revenant* (2016), humans have ultimately won. Wilderness, especially in the West, is greatly under threat, and can survive only with the protection of willing governments through the creation of national parks and conservation zones in which human presence is limited. The control and rationalization of wild places is a difficult contradiction to manage; the provision of opportunities for unfettered exploration and solitude has to be balanced against conservation imperatives and rising visitor numbers, as growing numbers of late-moderns seek escape from the everyday. Overcrowding can lead to what Hall and Shelby (2000) term 'displacement', where the dissatisfied choose to visit less crowded places (spatial displacement) at less crowded times (temporal displacement). Instances of displacement can be seen in climbing tourism; in the case of Li Ming, the climbers' relationship with the concept of the national park is problematic.

Li Ming belongs to the recently formed (2009) Mount Laojun Mountain National Geological Park, which, amidst outstanding mountain scenery, contains 171 species of endangered birds and mammals that are named on the Convention on International Trade in Endangered Species (CITES) list. One of these is the highly endangered Yunnan golden monkey, of which only 2,000 remain in the wild. The rapid development of national parks in China has inspired debate on their effectiveness and purpose, and there are concerns about whether they commodify wilderness as much as conserve it (Wang *et al.* 2012). Discussions on the many forms of the commodification of wilderness and nature often centre on the question of park fees, and the feelings of visitors towards paying them. Trainor and Norgaard (1999) found discontent among visitors to US parks about the fees, but an overall willingness to pay them to gain entry and experience wilderness. Their research explored the possibility of simultaneous pragmatic support for wilderness fees and conceptual disapproval of treating the wilderness as a commodity. In Li Ming, the compulsory park fees meet with resistance from some climbers. During conversations before and during my visit, I was frequently told that payment could be avoided by arriving late at night when the park office was closed. One climber explained that she had made numerous visits to the park in recent months and on every occasion she had evaded payment by taking the late bus from Lijiang and waiting for several hours until the office staff had left. Dirtbags are known for their thriftiness in order to maximize their climbing time and delay their return to paid work. Mount Laojun's entrance fee is 100 yuan (approximately £11.50 or US$14 at the time of writing), which another fee-evading climber equated to two days' living expenses in China – an ample excuse to avoid paying it, as far as he was concerned.

Beyond financial issues, though, climbers also expressed unease with the rationalization of the Li Ming area, which contains many waymarked trails and some large-scale, visually obtrusive signage, unsympathetic landscaping – concrete steps with steel hand-rails, for example – and a hilltop cable car that is visible from many of the surrounding cliffs and peaks. These features, regarded as anathema to wilderness, generated both amusement and resentment while also symbolizing the distinction between the climbers and 'tourists'. There is tension in climbing discourse between the perceived difference in desire and need for facilities between climbers and non-climbers; as discussed above, dirtbags prefer their provisions to be basic, natural and cheap. There was general derision of the conspicuous signage, artificial walkways and other evidence of the rationalization of this wilderness, no more so than towards the Thousand Turtle Mountain cable car, situated just 500 metres from Li Ming village. Users of this service were denigrated as 'fat', 'lazy' and – seemingly worst of all – 'tourists'.

China's domestic tourism market is exploding, bringing investment for infrastructure and facilities development to its areas of natural beauty. Li Ming's expansion was visible even during the single month when I was present, with a constant stream of heavy construction traffic rumbling through the narrow, cobbled main street towards the upper reaches of the valley, where new luxury hotels were being built. The growing influx of day-trippers and holidaymakers from urban China necessitates the provision of car parking, restaurants, public toilets and more hotels to cater for the country's burgeoning leisure class. This contradicts the romantic vision engendered in Dobie's guidebook, which states, 'we invite you to explore this undeveloped part of the world, experience the climbs, and discover what makes Li Ming truly unique' (Dobie 2015: 8). In some ways, the climbers' resistance to this resembles that of the 'preservationist', who, in the context of North American national park usage, 'sees mountain climbing as self-reliance, for example, whereas climbing in an electrified tramway is perceived as a passive and dependent activity' (Sax 1980: 14). Self-reliance is a quality of self-making that is threatened if adventures are overly facilitated, since finding one's own way, whether by navigation or hard uphill slog, defines the adventure at this end of the continuum. There is an expectation among climbing tourists in a place that is remote from Europe and the United States and offers non-rationalized and adventurous climbing that it should not resemble a 'tourist trap' or, as one climber put it, a 'sandstone theme park'. Such expectations are evident in Dobie's guide, which warns about climbing in the developing world and some of its concomitant dangers:

> Li Ming and Laojunshan [*sic*] Park are located in a developing part of the world. It is up to you to keep yourself and your climbing party safe. Climbing is an inherently dangerous activity. In the event that an injury requiring medical attention occurs, the closest medical facility is located roughly three hours away in Lijiang, Yunnan. There are no helicopters or rescue services here. Please take extra caution when climbing in the park. This is a wilderness setting and a fun day out could easily turn into a rescue situation.
>
> (Dobie 2015: 2)

Urry's (2002) 'post-tourists' are understood to be aware of the staged nature of the experiences of tourism, drawn to the spectacle and other attractions of commodified place-related tourism yet still prepared to enjoy playing the role of tourists (Crick 1995; Cloke and Perkins 2002). Climbing tourists, on the other hand, expressly deny the role of tourist and denounce the commodification of their playgrounds (Bott 2015). Lewis (2000) discusses the potential for human embodied agency to be affirmed by climbing, distinguishing the climbing body from the metropolitan body. He suggests that the former reaches a state of freedom through, in part, its flirtations with death and danger. In the absence of mediation via formal discourse (of established climbing literature, for example), climbing destinations such as Li Ming gain reputations and fame in the discursive circles in which climbers manoeuvre (climbing blogs and websites, conversations, social media and so on). Thus, notions of ideal spaces run deeply and keep locations such as this on specific non-corporate 'radars'. Knowledge of places with what might be called 'underground' appeal is produced and circulated, and when the valued location is threatened by rationalization and control, a disjuncture is created. This connects to the idea that mountain adventure tourism is perhaps an impossible contradiction:

> To feed their own compulsion to climb, mountaineers have initiated and participated in various endeavours that attract an ever-growing number of people to mountain regions … led to a loss of meaning and value in the experience as mountaineers confront at each step in their travels the adverse impact of their presence … [T]heir mountain trek and summit struggles represent an ultimate challenge and respect for the 'wildness' of nature; yet as a consequence of their presence, they transform, tame, and degrade 'nature'.
>
> (Johnston and Edwards 1994: 473)

Yet it is a necessary contradiction in most circumstances. Climbers who travel abroad resemble any other tourist insofar as they need transport, food and shelter. Many also require internet access, retail opportunities and so on; and, as noted above, the discovery and occupation of a mountain environment immediately transforms and 'tames' it.

Cross-cultural encounters and the 'spectacle of the Other'

Remoteness is, to some degree, a personal sensation that involves an embodied feeling of being somewhere Other and differentiating, measured through landscapes and environments that are unfamiliar, distinct and possibly hostile. This is experienced in part through access to the rarefied nature of the climbing in Li Ming and the particular markers of cultural and social capital required to get there. In addition to the visual contours of the landscape, embodiment in culturally distinct environments is discursively important, and this was articulated in the comments and actions of the climbers. The local people were the focus of much discussion and photography, and featured heavily in the climbers' social media posts. Returning to Dobie's (2015) guidebook as a point of conception for climbers, the opening section devotes a notable degree of prominence to describing the local Lisu, Naxi, Yi and Tibetan people, alongside the usual logistical information about transport, shopping, maps, other attractions and the like. This is interesting, since guidebooks for other parts of Asia, such as Ton Sai in Thailand, do not normally include a section on the local residents. Dobie stresses the unique nature of Li Ming in terms of its geology and landscape, then turns his attention to 'minority groups', whose history and traditions are described and 'showcased' in photographs.

The positioning of Indigenous groups alongside other attractions is common practice in more mainstream tourism promotional literature, particularly in nascent markets for 'ethnic' or 'tribal tourism' in the Global South, where indigeneity is displayed as a tourist attraction in problematic ways because of its potential to 'encapsulate and control the other, to stop motion and time, to exert power' (Bruner 2005: 195). Indigeneity is often depicted as a signifier of unique, 'authentic' experienes as tourism promoters are well aware of consumers' growing desire for 'authenticity' and utilize processes of 'otherizing' and 'traditionalizing' to exploit and perpetuate that desire (Koontz 2010). This is problematic in many ways, not least because it exoticizes, reifies and exploits local communities, and the commoditization of difference implies that what were once personal cultural displays become cultural products, which may lead to the invention of heritage to meet visitor conceptions of the Other (Hall and Tucker 2004).

The inclusion of photographs and descriptions of local Indigenous communities in a climbing guide that emphasizes the area's uniquely adventurous nature suggests that indigeneity is enmeshed in thoughts and notions of adventure; that the distinction sought in niche climbing/travel practices continues through encounters with 'exotic' others. This was evident in the climbers' conversations, and in their photography and social media posts. The climbers seemed to feel it was important to present themselves as corporeally located in exciting environments, not simply in terms of rock-climbing but also in relation to 'exotic'/foreign/strange food (deep-fried crickets, insect pupae, etc.) against a backdrop of 'local colour' provided by the Indigenous people (Morgan and Pritchard 1998). Indigeneity is being packaged and commodified alsongside adventures for 'white' consumption by Western climbers, where ethnic people are 'invoked as natives' (Desmond 1997). The pushing of frontiers in climbing tourism by heroic Westerners is reminiscent of the symbolism of colonial exploration, where muscular individuality and independence overcome adversity and beget true adventure, and the presence of Others brings the self into relief.

Conclusions

Rural China, like many other areas in the Global South, occupies a special place in the imagination of Western adventurers, often figuring discursively as remote, faraway, different and exotic, and with satisfyingly little in the way of safety infrastructure, rescue services, rationalization and so on. The slower pace of development means that numerous difficulties in terms of language, transportation, facilities and communication must be overcome and this is done with relish by those seeking to distinguish themselves from the ease and routine of mass tourism to more accessible and touristic places, or even adventure tourism to 'safer' areas. These points of tangible adversity and extremes make adventure tourism to developing areas especially desirable, because both the activity and the environment are risky, creating adventures within adventures, where various 'selves' may truly come to life.

But this is neither straightforward nor unproblematic. The presence of tourists (including climbers) in wilderness inevitably precipitates change and interference, which disrupt that deeply felt longing for the wild. There is tension between idealistic notions of what individuals expect to find and the realities that emerge, even in the world's highest mountain ranges. The 'East' still bears the Orientalist fantasies of Western tourists seeking out the exotic, primitive and undeveloped, which engenders unrealistic expectations about the wilderness and more broadly the paradoxical fantasy of uninhabited tourist space. As Meethan (2001: 37) suggests: '[S]truggles over the symbolic construction of space are struggles to objectify meanings, to impose on, or appropriate from the environment a particular order, a dynamic process of contestation and appropriation through which particular interests are maintained and legitimised.'

Thinking through the objectification of meanings of this particular context brings us back to this chapter's opening epigraph. In it, Cronan suggests that the wilderness is a mirror that reflects only our own longing and desires; it has no true essence in itself. Therefore, our quest for wilderness becomes purely narcissistic. Our expectations and appropriations stem largely from historical and literary visions of the uncontrolled violence of the western frontier that are so deeply embedded in the Western imaginary, and from our basic definitional understanding of what it is to be wild: natural, independent of humans, difficult to access, unpredictable, untamed and certainly non-rationalized. There is well-established contestation and conflict between the opposing interests of wilderness-seekers and state/park authorities in the Global North, yet visitors to 'Far Eastern' wilderness areas seemingly do not expect to encounter similar issues there. Theirs is an intersection of desires for pre-capitalist, primitive, tourist-free adventures. When climbing is added to the mix, the interests and appropriations become trifold: those of tourist seeking the exotic; those of adventurer seeking wilderness; and those of climber seeking frontier climbing. As this chapter has attempted to illustrate, while some of those expectations are met by Li Ming, others are not, creating some discontent, frustration and potentially an unfavourable reflection of the self-as-tourist.

References

Bauman, Z. (1998) *Globalisation: The Human Consequences.* Cambridge: Polity Press.

Beedie, P. (2003) 'Emergence of Mountain-based Adventure Tourism', *Annals of Tourism Research*, 30(3): 625–643.

Bott, E. (2009) 'Big Mountain, Big Name: Globalized Relations of Risk in Himalayan Mountaineering', *Journal of Tourism and Cultural Change*, 7(4): 287–301.

Bott, E. (2013) 'New Heights in Climbing and Tourism: Jordan's Wadi Rum', *Journal of Tourism and Cultural Change*, 11(1–2): 21–34.

Bott, E. (2015) '"You Can Never Cross the Same River Twice": Climbers' Embodied Quests for "Original Adventure" in Southern Thailand', *Tourist Studies*, 15(1): 101–116.

Bourdieu, P. (1984) *Distinction: A Social Critique of the Judgement of Taste.* London and New York: Routledge & Kegan Paul.

Bruner, E.M. (2005) *Culture on Tour.* Chicago: University of Chicago Press.

Cloke, P. and Perkins, H. (2002) 'Commodification and Adventure in New Zealand Tourism', *Current Issues in Tourism*, 5(6): 521–549.

Cohen, E. (1988) 'Authenticity and Commoditization in Tourism', *Annals of Tourism Research*, 15(3): 371–386.

Crick, M. (1995) 'The Anthropologist as Tourist: An Identity in Question'. In: M. Lanfant, J. Allcock and M. Bruner (eds) *International Tourism: Identity and Change.* London: Sage, pp. 205–223.

Cronan, W. (1996) 'The Trouble with Wilderness: Or, Getting Back to the Wrong Nature', *Environmental History*, 1(1): 7–28.

Desmond, J.C. (1997) 'Invoking "The Native": Body Politics in Contemporary Hawaiian Tourist Shows', *Drama Review*, 4(4): 83–109.

Dickey, J. (2013) 'The Climbers' Republic of Li Ming, China', *Climbing*, 8 April. Online. Available at: www.climbing.com/places/the-climbers-republic-of-liming-china/. Accessed 27 April 2017.

Dobie, M. (2015) *Li Ming Rock.* Self-published. Available by email request from the author.

Douglas, N., Douglas, N. and Derret, R. (eds) (2001) *Special Interest Tourism: Context and Cases.* New York: John Wiley and Sons.

Hall, C.M. and Tucker, H. (2004) *Tourism and Postcolonialism: Contested Discourses, Identities and Representations.* London: Routledge.

Hall, T. and Shelby, B. (2000) 'Temporal and Spatial Displacement: Evidence from a High-use Reservoir and Alternate Sites', *Journal of Leisure Research*, 32: 435–456.

Johnston, B.R. and Edwards, T. (1994) 'The Commodification of Mountaineering', *Annals of Tourism Research*, 21(3): 459–478.

Koontz, A. (2010) 'Constructing Authenticity: A Review of Trends and Influences in the Process of Authentication in Consumption', *Sociology Compass*, 4(11): 977–988.

Lennon, J. and Foley, M. (2002) *Dark Tourism: The Attraction of Death and Disaster.* London: Continuum.

Lewis, N. (2000) 'The Climbing Body, Nature and the Experience of Modernity', *Body and Society*, 6(1): 58–80.

Meethan, K. (2001) *Tourism in Global Society: Place, Culture, Consumption.* London: Palgrave.

Morgan, N. and Pritchard, A. (1998) *Tourism Promotion and Power: Creating Images, Creating Identities.* Chichester: Wiley.

Mortlock, C. (1994) *The Adventure Alternative.* Milnthorpe: Cicerone Press.

Mowforth, M. and Munt, I. (2009) *Tourism and Sustainability: Development, Globalization and New Tourism in the Third World*, 3rd Edition. London: Routledge.

Nerlich, M. (1997) *Ideology of Adventure: Studies in Modern Consciousness.* Minneapolis: University of Minnesota Press.

Novelli, M. (ed.) (2005) *Niche Tourism: Contemporary Issues, Trends and Cases.* Oxford: Elsevier Butterworth-Heinemann.

Palmer, C. (2004) 'Death, Danger and the Selling of Risk in Adventure Sports'. In: B. Wheaton (Ed.) *Lifestyle Sport: Consumption, Identity and Difference.* London: Routledge, pp. 55–56.

Priest, S. (1990) 'The Adventure Experience Paradigm'. In: J.C. Miles and S. Priest (eds) *Adventure Recreation.* State College, PA: Venture Publishing, pp. 157–162.

Rhinehart, R. (2003) *To the Extreme: Alternative Sports Inside and Out.* New York: State University Press.

Rickly-Boyd, J. (2012) 'Lifestyle Climbing: Towards Existential Authenticity', *Journal of Sport and Tourism*, 17(2): 85–104.

Robinson, M. (2005) 'Foreword'. In: M. Novelli (ed.) *Niche Tourism: Contemporary Issues, Trends and Cases.* Oxford: Elsevier Butterworth-Heinemann, p. xix.

Sax, J.L. (1980) *Mountains without Handrails: Reflections on the National Parks.* Ann Arbor: University of Michigan Press.

Sharpley, R. and Stone, P. (eds) (2009) *The Darker Side of Travel: The Theory and Practice of Dark Tourism.* Bristol: Channel View Publications.

Simmel, G. (1965) 'The Adventure'. In: K. Wolff (ed.) *Essays on Sociology, Philosophy and Aesthetics by Georg Simmel.* New York: Harper and Row, pp. 243–258.

Stebbins, R.A. (2002) *Organizational Basis of Leisure Participation: A Motivational Exploration.* State College, PA: Venture Publishing.

Swarbrooke, J., Beard, C., Leckie, S. and Pomfret, G. (2003) *Adventure Tourism: The New Frontier.* Oxford: Elsevier.

Timothy, D.J. (2002) *Tourism and Political Boundaries.* London: Routledge.

Trainor, S.F. and Norgaard, R.B. (1999) 'Recreation Fees in the Context of Wilderness Values', *Journal of Park and Recreation Administration*, 17(3): 100–115.

Urry, J. (2002) *The Tourist Gaze*, 2nd Edition. London: Sage.

Varley, P. (2006) 'Confecting Adventure and Playing with Meaning: The Adventure Commodification Continuum', *Journal of Sport and Tourism*, 11(2): 173–194.

Wang, G., Innes, J.L., Wu, S.W., Krzyzanowski, J., Yin, Y., Dai, S., Zhang, X. and Liu, S. (2012) 'National Park Development in China: Conservation or Commercialization?', *Ambio*, 41(3): 247–261.

12 On the trail of the new frontier

Doing things the hard way in Australian overland travel

Rosemary Kerr

Introduction: the fabled 'frontier'

In Jack Kerouac's *On the Road* (Kerouac 1972: 230), Dean Moriarty exclaims to Sal Paradise, 'Man, the road must eventually lead to the whole world. Ain't nowhere else it can go – right?' Kerouac's semi-autobiographical novel, first published in 1957, became a defining classic of the post-war American road genre, epitomizing the zeitgeist of America's love of mobility and the freedom of the open road. Yet, as the characters drive and hitchhike across the country from east to west and back in modern automobiles, they seek physical, mental and spiritual rejuvenation through contact with a more authentic, rural, pre-industrial, or 'frontier' American landscape and people (Shaffer 2001).

The central – even sacred – place of the road in the American national psyche is often explained by the unique influence of the 'frontier' in American culture. Indeed, the 'frontier legend' is one of America's most powerful and enduring national mythologies. Historian Frederick Jackson Turner's landmark 'frontier thesis' of 1893 argued that the constant movement of Americans westward, expanding settlement across a rapidly diminishing 'frontier' – the place where Americans struggled to master the continent – has been the greatest influence shaping America's national character, producing a practical egalitarian people with a democratic outlook, based on the freedom and individualism that land ownership allowed (Grossman 1994; Faragher 1998). As the physical frontier was closing by the late nineteenth century, Americans are said to have embraced the idea of the open road as a symbol of freedom, movement, progress and opportunity, and a means by which the frontier could be continually recreated and rediscovered (Patton 1986; Primeau 1996; Jakle and Sculle 2008). The 'frontier' was redefined as a 'zone of open opportunity'; a state of being 'on the edge of exciting possibilities'; a metaphor for 'promise, progress and ingenuity' (Limerick 1994: 67–68; Kerr 2013).

Frontier mythologies and landscapes are also influential in Australian culture. The Australian 'frontier', however, was very different to that of America. As Ward (1977), White (2009) and Clarsen (2010b) have noted, the latter was characterized by small-scale agricultural holdings and it had been relatively densely and rapidly settled by the late nineteenth century. In contrast, much of the Australian continent, particularly the interior or 'outback', remains sparsely settled due to the arid conditions, with large-scale pastoralism dominating both economically and culturally.

Central to the Australian 'bush' and 'pioneer legends' is the idea of movement across a 'frontier' environment in shaping the character of a nation and its people. Such legends were created in the 1880s and 1890s as technological advances in communications and transport, including the telegraph and railways, lessened the traditional divide between the city and the bush (Waterhouse 2005). Artists, writers and poets such as A.B. 'Banjo' Paterson and Henry Lawson, nostalgic for what they perceived was a disappearing frontier society, attempted to define a distinctive Australian culture, inspired by the growth of nationalism and a search for new heroes and symbols. According to Russel Ward's *The Australian Legend*, first published in 1958, it was the tough, resourceful, 'nomad tribe' of itinerant pastoral workers of the nineteenth century – the shearers, drovers, boundary-riders and rouseabouts[1] – who came to represent most powerfully a distinctive Australian type, embodying qualities of courage, resilience, pragmatism, egalitarianism and a love of freedom and wide open spaces (Ward 1977). The 'bush' legend is epitomized by Paterson's drover, 'Clancy of the Overflow' (1889) or his great bush horseman, 'The Man From Snowy River' (1890).

Historian John Hirst (1978: 316–319) identified the related 'pioneer legend', which also emerged in the art and literature of the same period. Emphasizing individual rather than collective enterprise, it celebrated 'those who first settled the land, as pastoralists or farmers' as well as the explorers who preceded them, braving hardships, 'subduing the land and battling the elements'. Paterson's poem 'The Pioneers' (1896), for example, is a nostalgic ode to 'the brave old pioneers … who rode the trackless bush in heat and storm and drought', forcing their way 'by tangled scrub and forests grim towards the unknown west' (Paterson 1917: 59–60). Hirst argued that in the absence of great warrior figures in Australia, explorers were seen as heroic for their bravery, often laying down their lives in their quest to blaze trails for others to follow. The pioneer legend was particularly broad and influential. These 'frontier legends' continue to resonate in the Australian cultural consciousness, as evidenced by films such as the *Crocodile Dundee* series (1986, 1988, 2001) and Baz Luhrmann's *Australia* (2008); and in the popularity of tourism sites such as the Stockman's Hall of Fame in Longreach, Queensland, which opened during the Australian bicentennial in 1988 (Smith 2012).

The Anzac legend, born during the First World War, became Australia's most powerful national foundation mytho. 'Anzac' is an acronym of the Australian and New Zealand Army Corps. Their involvement in the ill-fated Gallipoli campaign on the Turkish peninsula in April 1915 is widely regarded as representing Australia's coming of age as a nation.[2] Journalist and war historian C.E.W. (Charles) Bean was instrumental in helping to create the legend. He selected stories, poems and images for his *Anzac Book* and multivolume *Official History* that demonstrated the same dogged courage he had identified in the Australian bush when writing *On the Wool Track* in 1910 (Broomham 2005: 1–3). Thus, the Anzac legend, with its defining national character traits of courage, endurance, sacrifice and mateship, grew out of the earlier bush and pioneer legends (White 1992: 130–131).

The legends also reflect Australia's history as a settler colonial society. As Johnston and Lawson (2000), Veracini (2010) and others argue, among the legacies of colonialism are anxieties and uncertainties over claims to place, which are resolved through constructing progressive, heroic narratives of origin to indigenize or naturalize the settler society and expunge the memory of Indigenous presence, and of the often violent practices by which Indigenous peoples were displaced. Outback or frontier landscapes are key sites in this indigenization mythology. Settler colonialism is an ongoing process and is manifested in cultural practices and representations, including travel and tourism.

This chapter explores the changing meanings and locations of the 'frontier' within broader historical and cultural contexts. From early transcontinental cyclists and motorists, to those who followed the 'hippie trail' across Asia and the Middle East to Europe in the 1960s and 1970s, to the recent phenomenon of Kokoda Track trekking, overland travellers have sought increasingly challenging adventures on and off the road, in Australia and beyond. 'Doing things the hard way' enabled Australians to recreate and perpetuate 'frontier' mythologies at home and abroad in negotiating self and national identity. For Australians, too, 'the road' really does lead to the whole world.

Hitting the road

The creation of Australia's frontier legends coincided with major revolutions in overland transport in the late nineteenth and early twentieth centuries: the bicycle and the motor car. Epic transcontinental journeys captured the nation's imagination. The adventurer Francis Birtles made several expeditions by bicycle and car, including two complete circuits of Australia and seven transcontinental crossings, between 1905 and 1912 (Birtles 1979). The title of the book that records his exploits, *Battle Fronts of Outback* (Birtles 1935), sums up his approach to travelling through the harsh Australian landscape.

Accounts of long-distance motoring journeys before and immediately after the First World War promoted particular car manufacturers and related accessories. The first motor vehicles were open, with no windscreens, roofs or doors, leaving motorists exposed to the elements. As roads remained relatively undeveloped and unsealed, dust and bogs were constant menaces, forcing travellers to negotiate intimately and directly with the surrounding landscape. These rough conditions suited the car makers' agendas as their vehicles' ability to survive long journeys in the toughest environments served to enhance their prospects of winning over potential customers. They also allowed pioneering motorists with a sense of adventure to demonstrate their resourcefulness and skill in conquering the landscape, via the new technology of the motor car (Kerr forthcoming).

Henry Dutton and his mechanic, Murray Aunger, were the first to cross the continent by car on their second attempt in 1908 (Figure 12.1). They drove from Adelaide to Darwin in a 25-horsepower Clement Talbot. In an interview for *Australian Motorist* magazine in September 1908, Aunger commented:

> Motoring is strenuous work when you have not only to find your course through a wilderness, but make a road for your car whilst advancing, and even build your own bridges. Crossing the Warrender, we had to make a corduroy track of branches and saplings, which we cut out of the scrub, for in the bottom of the creek was a treacherous quicksand.
>
> (*Australian Motorist* 1908: 17)

These remarks reveal that the motorists viewed themselves as true pioneers, literally charting a course and taming the frontier (Kerr 2013).

MR AUNGER DRIVING THROUGH A RIVER VALLEY

Figure 12.1 Henry Dutton and Murray Aunger drive through a river valley in central Australia during their first (unsuccessful) attempt to cross the continent by car in 1907

Photograph courtesy of H. Dutton Collection, Northern Territory Library [Image 10070/12008]

The First World War was a defining period in the development of motor-vehicle technology and the formation of the Australian national identity. During the war, motor vehicles convincingly proved their worth for long-distance transport. By the 1920s, cars had improved and were increasingly accessible to a larger proportion of the population, rather than the preserve of the upper classes (Lee 2010: 235–236; Kerr forthcoming). Clarsen (2008) reveals that women were instrumental in this democratization process, as their enthusiasm for motoring encouraged manufacturers to make cars easier to drive.

Clarsen (1999, 2008) also argues that women's around-Australia motoring journeys in the 1920s played a powerful role in the process of 'peacefully recolonizing' Australia by car. Between 1925 and 1927, three pairs of female motorists set out to make a circuit of the continent. The first to succeed was Mrs Marion Bell, who left Perth with her eleven-year-old daughter (also called Marion) in October 1925, travelled in a north-easterly direction across the top of the continent, continued on to Brisbane, Sydney and Melbourne, and arrived in Adelaide in April 1926. The Bells were followed by Gladys Sandford and Stella Christie from New Zealand in March 1927, and by two young society women, Jean Robertson and Kathleen Howell, later that year. The press and public followed the progress of all three expeditions with great interest and fanfare. News reports and sponsors' advertisements marvelled at the women's 'pluck', adventurous spirit and tenacity in overcoming obstacles, difficulties and 'gruelling tests', including 'long stretches of parched, arid wastes; shifting sand; treacherous rivers and creeks … undefined and rugged tracks' (Clarsen 2008: 138).

Even more importantly, the journeys proved that, if these women could do it, anyone could, as long as they had access to the superior technology of the motor vehicle. Bishop (1998) and Clarsen (2008) argue that the act of completing a circuit of the nation, driving as close as possible to the outer limits of the frontier, symbolically patrolled Australia's borders and laid claim to all the land within. Clarsen (2008) also suggests that the burgeoning interest in around-Australia and transcontinental motoring from the 1920s onwards was another means of linking disparate geographical regions and states into the unified, coherent idea of a 'nation' in the aftermath of federation and the First World War.

A quest for the 'real' Australia

From the 1920s onwards, *Australian Motorist*, the first national motoring journal, founded in 1908, began to promote long-distance outback motoring to a wider audience. Seeing and knowing a so-called 'real Australia' – defined by well-established images of the 'bush' and 'outback' – became an important cultural preoccupation in the interwar years and into the mid-twentieth century. During this period, Australia was becoming increasingly urbanized and industrialized as well as 'nationalized'. Ever more people moved from the countryside to the cities in search of work and opportunities in factories and industries (Waterhouse 2005). This transformation in social and economic life challenged long-held beliefs about the national character and was often manifested in a nostalgic, romanticized vision of rural, and particularly outback, Australia, which was celebrated by artists, writers and intellectuals as an antithesis to the cities and suburbs.

This cultural response to urbanization and modernization was not unique to Australia. As Russel Ward (1978: 190) has observed: 'From ancient Rome men [*sic*] have been impelled to romanticize and identify with "outback" or "frontier" virtues, whenever they have begun to be herded into cities.' As Shaffer (2001) and Davis (2008) point out, the development of 'romantic nationalisms' was a trend that was evident throughout the developed world, but particularly in the United States, Britain and Northern Europe. In America, the 'West' was further mythologized in the 1920s as Americans experienced increasing anxiety towards the forces of industrialization, incorporation and urbanization that were transforming North America into a complex, modern nation (Shaffer 2001). For both Australia and America, the 'real' heart, soul and spirit of the nation was to be found in a romanticized region, far from the well-known cities: for Americans, it was the West; for Australians, the centre or outback (Kerr forthcoming). Phrases such as 'Red Centre' and 'Red Heart' associated with central Australia and the iconic Uluru (Ayers Rock) – a place long sacred to Aboriginal peoples – attest to the region's significance in the broader national psyche.[3]

Descriptive travel writing was one of the most popular literary forms in Australia between the wars, and its popularity continued after the Second World War (Bonnin, 1980; Fetherstonhaugh, 2002; see also Chapter 2, this volume). Writers such as Ernestine Hill, Frank Clune, Ion Idriess, Arthur Upfield and George Farwell did much to promote and romanticize the Australian outback. *Walkabout* magazine, published by the Australian National Travel Association (now Tourism Australia) between 1934 and 1974, declared that its editorial policy was 'an educational crusade' to enable Australians and people of other lands to 'learn more of the romantic Australia that exists beyond the cities' (Ewers 1959: 44).

Narratives by established travel writers as well as ordinary motorists reveal the importance of travelling overland so that they could really engage with the landscape and meet 'real' Australians. Leslie and Coralie Rees explained that they chose to hitchhike through north-western Australia, rather than travel by rail or air, because 'We wanted to do things the hard way … We wanted to see the land as the pioneers saw it, and to meet the people who belonged to the land and were in some instances still pioneers' (Rees and Rees 1953: 14). Overland travel was promoted as the best way of truly seeing, understanding and appreciating Australia, and it enabled travellers to enhance their own Australian identity.

It is worth noting that even in contemporary Australian society, the trope of the 'pioneer' remains pervasive. Laurajane Smith's (2012) research into visitors' responses to the Stockman's Hall of Fame in outback Queensland reveals that many consider the 'effort' entailed in travelling to the hall – which involves driving over long distances and witnessing the changing landscape – as an essential part of the experience. This 'secular pilgrimage' to the outback, which many still regard as the 'real' or 'more authentic Australia, strengthens their own understanding of and identification with the pioneer legend celebrated in the hall's exhibits. Similarly, a recent tourism initiative in South Australia – the Great Australian Outback Cattle Drive – held periodically since 2002, allows city dwellers to experience droving life along some of the nation's legendary stock routes: the Birdsville Track and the Oodnadatta Track. Tourism Australia promotes the event as a chance to 'trace the trail of Australian pioneers who forged their way through these raw, powerful landscapes … Meet modern-day droving legends and feel like one yourself' (Tourism Australia 2017; Kerr forthcoming).

Following in the footsteps of explorers and pioneers

For the earliest cyclists and motorists, it was easy to identify as pioneers, but from the 1920s and 1930s onwards, motor vehicles became more common and roads improved, making journeys more comfortable. As more travellers ventured into remote regions, it was also harder to be the 'first' or 'only' pathfinders. Later travellers, from caravanners to four-wheel-drive enthusiasts, 'grey nomads', backpackers and even long-distance truck drivers, sought ways of recasting their own journeys as pioneering feats, recapturing the spirit of adventure and enabling the 'frontier' to be recreated continually (Fetherstonhaugh 2002; Hall 2006; Kerr 2012, 2013 and forthcoming; White 1997, 2005). Pioneering imagery was often invoked in promoting new modes of transport. Caravans, for example, which gained popularity in the 1930s, appealed to those with the 'pioneering spirit' who sought 'new horizons' (Australian Motorist 1937b: 154).

As Eric Leed (1991) notes, the very nature of travel – particularly 'heroic' travel – has long been considered as an experience in which suffering, hardship, ordeals and danger are essential components, testing and refining the character of the traveller. The sense of adventure and reward for effort achieved by 'doing things the hard way' heightened the appeal of long-distance road travel for many. Muriel Dorney (1928: 188), on her 'adventurous honeymoon' around Australia in 1928, declared that 'what proved to be the most dreary and monotonous portion of the whole trip' was when 'the track was good most of the time' and she and her husband 'had neither sand nor rivers with which to contend'. After completing a circuit of Australia in 1932, William Hatfield (1936: 44) remarked: 'With a Roman road from Sydney to Darwin there would have been no adventure in taking a Minx up there.' The narrowness of Hatfield's car meant that 'roads' consisting of tracks made by other vehicles were of little use to him. This enabled him to identify as a pioneer, as he wrote, 'With one wheel, at any rate, I would have to blaze a trail' (Hatfield 1936: 5).

From the 1930s to the 1950s, travellers' narratives reflect a fascination with nineteenth-century explorers' journeys (Fetherstonhaugh 2002; White 2009). Many travellers read these early explorers' accounts in preparation for their own trips, following the same routes. Motorists usually paid homage to the explorers who blazed trails for them to follow, and often contrasted their own relatively easy journeys by motorized transport on well-defined roads with the hardships endured by the original pioneers. At times, however, closer identification was possible. Leslie Rees wanted to 'move over the old tracks of Alexander Forrest' (Rees and Rees 1953: 112), an explorer of the Kimberley region of Western Australia. When the truck in which Rees was travelling broke down in a creek along the Fitzroy River, he contemplated: 'Here, in the remote Kimberleys, one felt that for hundreds, even thousands, of square miles the scene since Forrest hadn't changed at all. Marooned in the creek, one might actually *be* Alexander Forrest' (Rees and Rees 1953: 138; original emphasis).

Following their hitchhiking travels, Leslie and Coralie Rees reflected: 'To share the experiences of the little knots of people spread over this vastness was emotionally enlarging' (Rees and Rees 1953: 282). For the couple, contact with representatives of a 'Northern way of life … tough, laconic, dryly casual, but quietly admirable in resource, pluck, endurance, and hard common sense' was personally transformative, enabling closer identification with the bush and pioneer legends.

Taming the frontier

'Discovering' the 'real' Australia through contact with a frontier environment and people has inspired strong emotional reactions in many travellers. Some experienced a kind of secular spirituality, inspired by the vast, ancient landscape. Sister Dora Burchill (1946: 11), a nurse who travelled with the mail truck to Innamincka in 1946, recorded that the '"everlasting sameness of the never-ending plain" does something to one's soul'. Biologists H.F. Broadbent and his wife, who drove through the north-west of Western Australia in the mid-1950s, experienced 'that feeling of awe and humility which affects man [*sic*] when gazing on scenes of natural grandeur. The feeling was enhanced by the silence and loneliness of the country we were in' (Broadbent 1955: 18–19).

In romanticizing the 'real Australia' of the remote outback and interior, a conflict emerged between revering its unique, awe-inspiring emptiness and isolation, on the one hand, and the desire to traverse and occupy that space, on the other. Australia's size relative to its population, its unprotected borders and its colonial past left a legacy of anxiety over white Australians' ability and entitlement to possess and develop the nation fully (Kerr forthcoming). While the fiction of a 'trackless bush' dismissed Aboriginal presence and provided a stage over which heroic pioneering trails could be blazed, it could also be troubling. As Reynolds (1998) and Curthoys (2007), among others, have shown, voices questioning the legitimacy of colonization and Aboriginal dispossession remained and 'this whispering in our hearts', as Reynolds calls it, grew louder over time. Such 'unsettling' permeates representations of Australian space, including the road (Falconer 2009; Kerr forthcoming). As discussed earlier, historians including Bishop (1998) and Clarsen (1999, 2008, 2010a, 2010b) see outback and around-Australia road tourism as an expression of white Australians' desire to cement ownership of every inch of the continent. The Redex Around-Australia Reliability Trials, which captured the public's imagination in the mid-1950s, resembled earlier motoring endurance tests, as entrants competed over increasingly arduous circuits off-road in remote regions, thereby conquering the 'untamed landscape through speed and strength' and once again demonstrating the 'pioneering spirit' (Clarsen 2010a: 361; Davidson and Spearritt 2000: 168).

In the years leading up to the Second World War, concerns over Australia's vulnerability to invasion and occupation by unspecified 'others' revealed underlying insecurities, at least partly informed by racial prejudice. William Hatfield (1936: 195) stressed the need for Australians to 'inhabit and exploit the vast North' before others did. An important part of making the region habitable for Europeans was to provide infrastructure, including roads. Hatfield saw the 'road-less and rail-less state' as forming an 'insuperable barricade on the frontier', favouring the invader and inhibiting its defence.

During and after the Second World War, the need to populate Australia's remote regions became more urgent. The Japanese advance and the bombing of Darwin in 1942 brought the prospect of foreign invasion even closer to home. The war also spurred development of strategic transcontinental roads, including the Stuart, Eyre and Barkley highways, which became more popular as tourist routes in the second half of the twentieth century.

Boring bitumen

More highways were bituminized during and after the war, becoming straighter, smoother and faster, and hence more comfortable to travel. While travellers generally appreciated the improved highways, many expressed nostalgia for earlier times, feeling that the country itself had been dulled as soon as it became easier to traverse. Kathleen Woodburn (1947: 16–17) lamented the change in the route that became the Stuart Highway as it was transformed from a 'track' to a 'road':

> Now we travel on gravel, levelled and coated with sand and bitumen. And yet, with the attainment of speed and comfort, we have lost something. The adventurous touch and the excitement of a hazardous enterprise will never return. The thousand miles from 'The Alice' to 'The Top End' is now a tourist's run, and often decried as monotonous … There is no spirit of adventure to keep one awake.

Commenting on his family's tour through the north-west of Western Australia in the mid-1950s, H.F. Broadbent (1955: 13–14) explained that they 'found the 800 miles of bitumen relatively dull and the 2,000 miles of dirt tracks exciting'. Bush tracks and rough roads allowed for greater engagement with nature, whereas they felt detached from the landscape on bitumen roads.

Road journeys are marked by psychological as well as physical 'boundaries'. For many travellers, their journey effectively concluded with the end of the most physically and mentally difficult phases. Edith Crake's narrative of her Nullarbor crossing on the Eyre Highway – which, in 1954, was only partially bituminized – ends just before Iron Knob, when she struck 'THE BITUMEN! Smooth and soft under the tyres. No corrugations, no potholes, no high mounds, and no real challenge' (Crake 1998: 61).

Off the road again

Australia's national highway system, which skirts the perimeter and bisects the continent, was developed and improved in the 1970s and fully sealed by the mid-1980s. As Hall (2005) argues, the blacktop effectively 'tamed' Australia, making transcontinental and around-Australia motor touring easier than ever. Yet, at this time, many travellers sought the kind of engagement with the landscape, adventure, achievement and exhilaration that had characterized the early years of bicycling and motoring by travelling off- rather than on the road.

Rex Ellis began conducting four-wheel-drive safari tours in the 1960s. One of his earliest trips was across the Nullarbor in 1966. Doing things the hard way by avoiding the highway or even the 'beaten track' was a priority. His tour company, Transcontinental Safaris, was the first to complete a tourist crossing of the Simpson Desert from Oodnadatta to Birdsville by Land-Rover in the mid-1970s. Ellis (1976: 81–82) described his clients, most of whom were from Melbourne, as having various motivations for making the journey, but they all shared a desire 'to travel across a region where no tourist had ever been before'.

While the Jeep became available to the general public after the war, four-wheel-drive tourism became increasingly popular in the 1980s, enabling adventurous motorists to explore the country 'off the beaten track'. Four-wheel-drives proved to be particularly popular with 'grey nomads' – retirees making extended round-Australia and transcontinental road trips – who were recognized as an emerging niche tourism market in the 1990s. Many of these nomads used their four-wheel-drives to tow caravans on sealed roads, then explored regions off-road in the car once they had established their base camps. In the era of mass tourism, travellers often sought a more personal and individualistic experience of 'discovery' and engagement (Bachman 1989; Bowden 2002). The growing popularity of adventure tourism and 'off the beaten track' destinations is a worldwide trend that has been explored in the tourism literature in recent years (e.g. Haines 2004; Laing and Frost 2014; Novelli 2005; Taylor *et al.* 2013).

Even as travellers sought to recapture the sense of adventure, challenge and discovery of earlier eras by venturing further off-road, there is evidence that the novelty was wearing thin for some by the late 1980s. In 1978, the Tamworth Four Wheel Drive Club placed a visitors' logbook at Poeppel Corner in the Simpson Desert, where the borders of Queensland, South Australia and the Northern Territory meet. Its entries, which continue up to 1987, reveal both appreciation of the technology that allowed the travellers to traverse the desert and disappointment that the experience was becoming all too easy or commonplace. In July 1987, one group wrote: 'Please, bigger sandhills required. Too easy! Not like the Sahara … All in all. Too Easy. Think we might go back the other way for a little hardship.'

The 'East' as the new frontier: overland to Europe

In the later twentieth century, the quest for adventure and doing things 'the hard way' began to extend beyond Australia. Despite – or perhaps because of – their isolation, Australians are often characterized as having a strong and long-held desire to see 'the road' as extending to a world beyond their borders. When Dutton and Aunger crossed the continent from Adelaide to Darwin in 1908, they followed the rough track of the Overland Telegraph Line, a vital communications link between Australia and Britain, which was completed in the early 1870s. Australian Motorist (1908: 19) saw the duo's feat as 'opening up possibilities of a grand motor car route linking Northern and Southern Australia … and a shortened route to Europe'.[4]

From the earliest days of motoring, there was great interest in how much of the world a motorist could travel overland with minimal reliance on boats. Throughout the 1920s and 1930s, there were several overland trips between England and Australia, including that of the intrepid Francis Birtles, who drove from London to Melbourne in 1928. Nine years later, Pioneer Tours proposed an International Overland Tour that would cover '25,000 miles by road', embracing five continents and twenty-five countries. Australian Motorist (1937a: 319) anticipated that the tour would 'mark a new development in overseas touring, [and] a new avenue of travel off the beaten track'.

Before the 1960s, however, most people travelled between Australia and Britain by ship. A notable exception was the famed Qantas 'Kangaroo Route', first flown from Sydney to London, with stopovers in Darwin, Singapore, Calcutta, Karachi, Cairo and Tripoli, on 1 December 1947 (Hubbard 2014). As Gerster (1995: 293–294) notes, the pilgrimage to Britain, which passed through the 'Far East' and the 'Near East', functioned as an Antipodean rite of passage. But, for most, the destination was far more important than the journey. Asia and the Middle East were mere stopovers en route to the 'real' destination – the 'centre' of civilization and culture in London and Europe. As a former British colony, Australia shared strong cultural and emotional, as well as economic and strategic, ties with Britain, and the idea of Britain as a 'spiritual home' was strong for many Australians of Anglo-Celtic descent, even into the late twentieth century.

Australia's relationship with Asia was more complicated. For more than a century, Australians defined Asia not as many diverse countries but as one region, and as a generalized threat (Broinowski 1992). From the 1960s, however, Australian attitudes to Asia and Britain began to change. When Britain entered into the European Common Market, Australian politicians, especially those on the Labor side, argued that the country needed to reorient itself towards Asia. For instance, Gough Whitlam, Prime Minister from 1972 to 1975, argued that Australians were 'Asians by an irrevocable act of geography' and should start thinking regionally (Gerster 1995: 294–295). The Vietnam War also focused attention on Asia as Australia's near neighbour.

For travellers, too, Asia was fast becoming a source of fascination rather than fear. Some younger Australians, impatient with their country's isolation, conservatism and reputation as a 'cultural desert', suddenly saw Asia as 'the place to be'. This shift coincided with a time in which Westerners, and especially young people, were increasingly attracted to the 'East', partly due to the influence of Beat writers such as Kerouac and the later counter-culture movement. In rebelling against the conservative mores of mainstream Western society, many young travellers turned towards Eastern cultures and traditions in search of alternative, more authentic ways of living and spiritual fulfilment. Travelling was also regarded as a path to self-discovery and transformation.

A new tourism phenomenon emerged in the 1960s and 1970s as waves of Australians, Britons, Europeans and Americans embarked on epic road trips that saw them driving, hitchhiking or taking local transport overland to and from Britain and Europe via South-East Asia, the Indian subcontinent and the Middle East. As Hampton (2013) points out, the 'hippie trail', as it was popularly known, also coincided with the growth of the hitchhiking movement in Europe, and was a prelude to modern backpacker tourism.

For Australians, the trail typically led through Indonesia, Malaysia, Thailand, India, Nepal – with its mecca of Kathmandu – and Afghanistan, then on to Iran, Turkey, Europe and finally London (Sobocinska 2006). Going overland was the cheapest way to travel. It also offered a more exciting alternative to sea or air travel through closer engagement with people and places along the way. Journalist and author Douglas Lockwood (1964: 14) likened air travel to a 'painless operation' in which the traveller 'might as well be anaesthetized or deep-frozen'. While motivations and experiences varied widely, many travellers embraced the opportunity that the overland route provided to revisit and perpetuate the pioneering ethos, so powerful in Australian culture, by 'doing things the hard way'.

Peter Jeans travelled by motorcycle from Perth to London in 1962–1963. He was inspired by a story he had read about an Englishman, Nick Lidstone, who had just completed the overland journey from London to Perth on a

Vespa scooter. Jeans (1998: xv) thought this was 'extraordinarily dangerous and adventurous'. The London–Sydney car rally of December 1968 is also credited with generating interest in travelling overland, and it certainly inspired Tony Wheeler (2005). Four years later, he made the overland expedition from London to Australia in a Morris van with his partner Maureen. Wheeler was adamant that he did not want to join one of the organized tours, which were becoming more prevalent around that time. Their guide to the trail, *Across Asia on the Cheap*, first published in 1973, claimed that travelling by local transport or hitching was 'the most genuine way of travel' as well as the cheapest (Wheeler 1973: 8). The Wheelers later founded the Lonely Planet publishing house.

Albert Laing (1975) shipped his Morris Mini Moke to Singapore, then Colombo, to commence his overland journey to London in the early 1970s. He considered only those who were driving their own vehicles to be genuine 'overlanders', while others were 'travellers', writing: 'The endeavour is more important than getting there, the journey is paramount and the arrival is incidental … I shall gladly turn my back on the easy life and cast my fate to the wind' (Laing 1975: 10–11).

To some extent, the 'East' became a new frontier for Westerners, but this frontier also retreated over time, as places along the trail were 'discovered' by ever more travellers who followed the first 'intrepids', as Rory Maclean (2006) called the trail-blazers of the 1960s. Places such as Bali, Kathmandu, Afghanistan, Iran and Turkey were transformed and modernized through contact with Westerners who wanted to experience the exoticism of remote, ancient civilizations and different ways of life. Western food, music and language all followed in their wake. Yet, the challenges of negotiating poor transport, chaotic traffic, ramshackle accommodation, lack of clean – and hot – water, illness and rough living conditions enabled even those travelling in the late 1970s to claim hardship as a badge of honour. Pauline Hyland (2009: 126), who took a bus tour from London to Kathmandu in 1977, described it as a 'trip that turned out to be heaven and hell and all that stuff in between. We earned our badges – other people buy them like airline tickets.' With regard to entering Iran, Hyland (2009: 28) commented that 'visiting a 4,000-year-old empire had to be worth tortuous travel'.

There is a sense that physical hardship allowed a closer connection or identification with the cultures visited. Trail travellers of all nationalities often linked the crossing of physical space with psychological and spiritual transformation. They were not only moving geographically, but temporally and imaginatively; stepping into the past, and into the 'East', seeking to leave their modern Western selves behind, to some extent at least. Jack Parkinson (2001: 17) promised that trail travellers would be 'reborn' or 'metamorphosed from the mundane to the exotic'. Peter Jeans (1998: 6) recorded the powerful impression of sighting Java Head in Indonesia, his first Asian stop and the first 'foreign landfall' of his life: 'I now reckoned myself to be a traveller … I was on my way.' Lieschen Muller, from Germany, described following the overland route as an 'Easternizing process' and suggested that going overland enhanced the feeling of gradual change (cited in Tomory 1998: 131).

Certainly, travellers often adopted the clothing – and entered into the customs and practices – of their host countries in their efforts to be indigenized. For most, the transition was temporary, but others experienced a more permanent transformation. One Australian traveller told an ABC radio documentary that he 'fell in love' with the architecture, landscape and people of Bali, eventually settled there, and 'set about becoming more Balinese than the Balinese' (O'Brien 2012). He was not alone. Trail travellers also influenced Australian culture. Tom Victor travelled to Nepal on the trail as a surfer in the 1960s. He returned in 1973 to complete a meditation course, then established Australia's first Tibetan Buddhist institute in Queensland in 1976 (Mackay 1976).

By 1979, parts of the overland route were effectively closed due to the Iranian Revolution and conflicts in Turkey, Afghanistan and Pakistan. Yet, the trail had a profound and lasting impact on Western culture generally, popularizing Eastern cuisine, philosophies and practices, such as yoga. It also influenced Australian tourism practices, fostering a fascination with South-East Asia as a destination that continues to this day.

A twenty-first-century frontier: trekking the Kokoda Track[5]

Since the 1990s, the phenomenal popularity of trekking the Kokoda Track in the mountains of Papua New Guinea has taken the link between physical and emotional frontiers to a new and extreme level. The track was a key Second World War battlefront, where dogged resistance by Australian forces in treacherous jungle terrain during 1942 halted the Japanese advance, turning the tide of the war in the Pacific. While pilgrimages to Kokoda by veterans and army cadets for commemorative occasions began in the 1950s, for years the campaign was virtually forgotten. In the late 1970s and throughout the 1980s, a number of trekking and adventure tourism companies began to offer small group treks over the track. For most Australians, however, it had little meaning and its story was overshadowed by Gallipoli.

Then, in April 1992, Prime Minister Paul Keating visited Kokoda for Anzac Day celebrations and to mark the fiftieth anniversary of the Kokoda campaign. Falling to his knees and kissing the 'sacred ground', he declared that if the Australian nation was founded at Gallipoli, then its 'depth and soul' were confirmed at Kokoda: there could be 'no deeper spiritual basis for the meaning of the Australian nation than the blood that was spilt [there]' (quoted in Scates 2013: 3). As Scates (2013: 233) points out, Keating wanted to shift the focus of Anzac memory away from Gallipoli, which he saw as an imperial conflict, and towards Kokoda, which was fought in defence of Australia. This coincided with his efforts to reorient Australia strategically and economically within the Asia-Pacific region, and with his vision for a more assertive, independent Australian identity. He drove campaigns for Australia to become a republic and to remove the Union Jack, which he saw as a reminder of the country's colonial past, from the national flag.

Between April 1992 and 1995, former army major Charlie Lynn – described as an 'adventurer' and a 'zealous patriot' – ran a series of commemorative treks to mark the fiftieth anniversary of the New Guinea campaign (Frith 1992b). Journalists, business leaders, sportspeople, politicians and veterans participated in these treks, which generated considerable publicity. Journalist Marion Frith (1992a) cited the opportunity for 'the adventure of a lifetime into one of the last frontiers' as an incentive to enrol.

Keating's successor as Prime Minister, John Howard (1996–2007), did much to rekindle enthusiasm for the Anzac legend, which boosted tourism to both Gallipoli and Kokoda. He sought to promote a reassuring, positive, unifying national narrative in response to the ascendance of what historian Geoffrey Blainey (1993) has called a 'Black Armband' view of Australian history, which Howard and other conservatives saw as overly pessimistic in emphasizing the negative impact of European colonization and its aftermath on Aboriginal peoples.[6] The so-called 'History Wars' engaged Australians in a public debate over the nature and control of national memory. Framed by conservatives as a battle between 'ordinary' Australians – justifiably proud inheritors of the pioneering legacy – and an intellectual, left-wing 'elite' who sought to discredit that legacy, the wars were fought in the mainstream media as conservative commentators, led by former academic and journalist Keith Windschuttle, challenged prominent historians' accounts of colonial frontier violence and massacres of Aboriginal people (Macintyre and Clark 2004).[7]

The real boom in Kokoda Track tourism came during Howard's leadership. During this time, documentaries, books, pilgrimages and memorials relating to both world wars proliferated, and many focused on Kokoda, facilitating what Jay Winter (2006) calls 'collective remembrance', which helped those who were not there to imagine.[8] Numbers grew from a few hundred permits issued in 2000 to a peak of nearly 6,000 in 2009, and the gruelling 96-kilometre course through heat, mud and steep jungle terrain continues to attract around 3,000 trekkers each year (Dart 2009; Olding 2012). Around thirty trekking companies now offer guided tours (Figure 12.2).

By following in the footsteps of the Australian soldiers, enduring hardship and suffering, and overcoming the physical and emotional obstacles of the track, trekkers actively seek to connect more closely with the memory and experience of the 'diggers', but also with a shared or national memory of the 'Anzac spirit' in addition to their personal and national heritage and identity. As the First World War has now passed from living memory, the Anzac legend is perpetuated by association with subsequent conflicts.

Trekking Kokoda is an intense physical, psychological and emotional experience that is *performed* in an environment that, ostensibly, has changed little since the war. Peter FitzSimons (2009a: 4) wrote: 'When you visit Gallipoli, you can see where they fought and imagine how they suffered; when you walk the Track you suffer a bit as they suffered.' Trekkers are aware that

Figure 12.2 Hiking the Kokoda Track, July 2009
Photograph by Luke Brindley, courtesy of Wikimedia Commons

they can never fully replicate the diggers' experience. Nevertheless, walking the track enables them to 'place' the diggers there and imagine what it was like for them; in doing so, they place themselves within that tradition. Trekking Kokoda is also a collective experience that allows mateship to be actively performed, as fitter, stronger group members help and encourage the weaker ones, thereby forming strong emotional bonds. Trekking is popular among corporate and sporting groups as a powerful leadership and team-building exercise.

Physical and mental suffering is an essential part of Kokoda trekking. As Scates (2013) points out, some express a desire for greater hardship and more rain, the better to replicate the diggers' experience in 1942 (an unusually wet year). Any efforts to make the track safer by widening it or building amenities for greater comfort meet with strong resistance. As Peter FitzSimons (2009b: 26) writes, 'A safer Kokoda Track just isn't the real thing.' By testing themselves against the challenges of the Kokoda Track, modern trekkers perhaps seek to prove themselves *worthy* of those who fought and died there, and as legitimate inheritors of broader Australian heritage and values. Suffering on the track also has a purifying, redemptive function. In recent years, those who have transgressed, or are considered 'outside' mainstream Australian society – such as a Lebanese youth who burned the Australian flag during the 2005 Cronulla race riots – have walked the track as a means of gaining readmission into a unified Australian culture. Finally, trekking Kokoda also reflects the changing way that current generations relate to the past and 'memory'. In recent years, both in the school curriculum and more generally, there has been a growing emphasis on family history and on placing personal stories within broader national narratives. As one trekker whose grandfather fought at Kokoda said, 'I feel more deeply in touch with my heritage … it's not just knowledge now, it's emotional. I know my country on a more personal level' (Jensen 2009: 3).

Conclusion

The quest for adventure, and the physical and emotional challenge of 'doing things the hard way', has been a recurring theme in Australian travel and tourism. 'Frontier' landscapes have played an important role in constructions of Australian identity at both an individual and the national level, and overland travel has a particular resonance in Australian culture. The foregoing discussion reveals that the 'frontier' is a space constructed as much in the imagination as by physical reality. The nature and location of the 'frontier' shifts with changing historical and cultural contexts. It is also a space that is often appropriated for political purposes. Frontier legends created in the late nineteenth century partly reflect a settler colonial society's need for heroic, progressive myths of origin. Identifying with and performing those legends is a way of taming unfamiliar, threatening environments – from the Australian outback to 'newly discovered' regions in Asia and the Pacific – as Australians have sought to negotiate and renegotiate their place in the nation and the world as well as their relationship to history.

The connection between emotion, affect and tourism was highlighted during the Romantic era, but more recently – together with the history of emotions generally[9] – has become the subject of greater scholarly interest (Picard and Robinson 2012). As Mike Robinson (2012: 24) points out, tourists' emotional responses to specific places are also socially and culturally conditioned. The resilience of the pioneer ethos and the quest for 'off the beaten track' tourism in outback regions perhaps reflects a nostalgic longing for connection with an affirming national mythology by a predominantly urban society, long disconnected from the 'bush'. Anxieties over the colonial past, which underlie Australia's foundation myths, remain unresolved and continue to surface. In the recent phenomenon of Kokoda Track tourism, the desire to identify with a national heritage born of heroic suffering has become stronger than ever, at a time when contested versions of the past are being publicly debated. The intense physical and emotional experience sought by Kokoda trekkers amplifies connection with the Anzac legend. It also reflects a growing trend towards experiential tourism. In traversing historic landscapes, modern tourists – and historians – increasingly ask not only 'What happened?' but 'What did it *feel* like?'

Notes

1 A rouseabout – or roustabout – was an unskilled general labourer.
2 The First World War was the first conflict in which Australian troops fought as a distinct unit – the First Australian Imperial Force (AIF), formed in August 1914 – rather than as part of the British Army.
3 Rosslyn Haynes (1998) explores the significance of the desert in Australia, including Aboriginal culture and representations in European literature, art and film over 200 years.
4 The Overland Telegraph Line crossed the centre of Australia from Port Augusta to Darwin, where it was connected by undersea cable to Java (Indonesia) and India, thence to Britain.
5 Much debate has surrounded the use of the terms 'track' and 'trail'. Most authors argue that the diggers themselves knew it as the 'Kokoda Track' and that 'trail' was an Americanism; hence, I use 'track' here.
6 Several developments in the 1990s brought the issue of Australia's treatment of Indigenous peoples to public attention. In his now-famous 'Redfern Speech', delivered at Redfern Park in Sydney on 10 December 1992 to launch the Year of the World's Indigenous Peoples, Paul Keating spoke frankly about the theft of Aboriginal land and children, and the impact of disease, alcohol, murder, violence and discrimination inflicted by the policies and practices of non-Indigenous Australians. This was the first time that a Commonwealth government had acknowledged the dispossession of Aboriginal and Torres Strait Islander peoples. The landmark High Court decisions of Mabo (1992–1993) and Wik (1996) recognized the existence of native title on Australian land. In 1997, the Human Rights and Equal Opportunity Commission published its report on the 'stolen generations', based on the commission's review of past governments' 'separation' policies, whereby Indigenous children were forcibly removed from their parents between 1910 and 1970.
7 Between 1994 and 2009, Windschuttle published several articles in journals such as *Quadrant* as well as a series of books on 'the fabrication of Australian history' with the aim of discrediting the work of historians such as Henry Reynolds and Lyndall Ryan on the extent of frontier violence and racism in Australia.
8 For example, Brune (2003), which sold 30,000 copies in six months, Ham (2004) and FitzSimons (2004).
9 The recent trend in historiography towards the history of emotions is exemplified by the establishment in 2011 of an Australian Research Council Centre of Excellence for the History of Emotions at the University of Western Australia. See: www.historyofemotions.org.au.

References

Australian Motorist (1908) 'A Trans-Australian Motorist: From Adelaide to Port Darwin: Mr H.M. Aunger Interviewed', September: 17–19.

Australian Motorist (1937a) 'International Overland Tour – 25,000 Miles by Road', February: 319.

Australian Motorist (1937b) Untitled article, November: 154.

Bachman, B. (1989) *Off the Road Again*. Melbourne: Lothian Publishing Company.

Bowden, T. (2002) *Penelope Bungles to Broome*. Sydney: Allen & Unwin.

Birtles, F. (1935) *Battle Fronts of Outback*. Sydney: Angus & Robertson.

Birtles, T.G. (1979) 'Birtles, Francis Edwin (1881–1941)'. In B. Nairn and G. Serle (eds) *Australian Dictionary of Biography*. Canberra: National Centre of Biography, Australian National University. Online. Available at: http://adb.anu.edu.au/biography/birtles-francis-edwin-5244/text8833. Accessed 2 May 2016.

Bishop, P. (1998) 'Driving around: The Unsettling of Australia', *Studies in Travel Writing*, 2: 144–163.

Blainey, G. (1993) 'Drawing up a Balance Sheet of Our History', *Quadrant*, July–August: 10–15.

Bonnin, M. (1980) 'A Study of Australian Descriptive Travel Writing, 1929–1945'. Ph. D. thesis, University of Queensland.

Broadbent, H.F. (1955) 'A Trip through the North-west', *Walkabout*, March: 18–19.

Broinowski, A. (1992) *The Yellow Lady: Australian Impressions of Asia*. Melbourne: Oxford University Press.

Broomham, R. (2005) 'Understanding the Place'. In: *ANZAC Memorial Hyde Park: Conservation Management Plan*. Unpublished report prepared for NSW Government Architect's Office.

Brune, P. (2003) *A Bastard of a Place*. Crows Nest: Allen & Unwin.

Burchill, D. (1946) 'Innamincka', *Walkabout*, April: 11.

Clarsen, G. (1999) 'Tracing the Outline of Nation: Circling Australia by Car', *Continuum: Journal of Media and Cultural Studies*, 13(3): 359–369.

Clarsen, G. (2008) *Eat My Dust: Early Women Motorists*. Baltimore, MD: Johns Hopkins University Press.

Clarsen, G. (2010a) 'Automobiles and Australian Modernization: The Redex Around-Australia Trials of the 1950s', *Australian Historical Studies*, 41(3): 352–368.

Clarsen, G. (2010b) 'Automobility "South of the West": Toward a Global Conversation'. In: G. Mom, P. Norton, G. Clarsen and G. Pirie (eds) *Mobility in History: Themes in Transport – T2M Yearbook 2011*. Neuchaet: Editions Aphil–Presses Universitaires Suisses, pp. 25–41.

Crake, E.M. (1998) *The Highway of Haunting Hilarity*. Perth, WA: the author.

Curthoys, A. (2007) 'The Volatility of Racism in Australia'. In K. Gelber and A. Stone (eds) *Hate Speech and Freedom of Speech in Australia*. Sydney: The Federation Press, pp. 20–33.

Dart, J. (2009) 'Business as Usual for Trekkers Ready to Go', *Sydney Morning Herald*, 13 August: 4.

Davidson, J. and Spearritt, P. (2000) *Holiday Business: Tourism in Australia since 1870*. Melbourne: Melbourne University Press.

Davis, T. (2008) 'The Rise and Decline of the American Parkway'. In C. Mausch and T. Zeller (eds) *The World beyond the Windshield: Roads and Landscapes in the United States and Europe*. Athens: Ohio University Press, pp. 35–58.

Dorney, M. (1928) *An Adventurous Honeymoon: The First Motor Honeymoon around Australia*. Brisbane: The Read Press.

Ellis, R. (1976) *Bush Safari*. Adelaide: Rigby.

Ewers, J.K. (1959) 'Discovering Australia through "Walkabout"', *Walkabout*, November: 44.

Falconer, D. (2009) '"The Poetry of the Earth is Never Dead": Australia's Road Writing', *Journal of the Association for the Study of Australian Literature*, July: 1–16. Online. Available at: http://openjournals.library.usyd.edu.au/index.php/JASAL/article/view/10156/10054. Accessed 26 September 2016.

Faragher, J.M. (1998) *Rereading Frederick Jackson Turner 'The Significance of the Frontier in American History' and Other Essays*. New Haven, CT: Yale University Press.

Fetherstonhaugh, T.J. (2002) 'The Journal, 'Walkabout' and Outback Australia 1930s–1950s: A Romantic Rapprochement with the Landscape in the Face of Modernity'. Ph.D. Thesis, Murdoch University.

FitzSimons, P. (2004) *Kokoda*. Sydney: Hodder Headline.

FitzSimons, P. (2009a) 'There are Few Bigger Challenges than the Track, but Its Pull is Irresistible', *Sydney Morning Herald*, 13 August: 1, 4.

FitzSimons, P. (2009b) 'And We Shall Remember Them', *Sydney Morning Herald*, 15 August: 26.

Frith, M. (1992a) 'Kokoda Trail Not as Tough as Ironing', *Canberra Times*, 20 October: 8.

Frith, M. (1992b) 'Hard Slog to Kokoda', *Canberra Times*, 15 November: 19.

Gerster, R. (1995) *Hotel Asia: An Anthology of Australian Literary Travelling to the East*. Ringwood: Penguin.

Grossman, J.R. (ed.) (1994) *The Frontier in American Culture: Essays by Richard White and Patricia Nelson Limerick*. Berkeley: University of California Press.

Haines, M. (2004) 'Off the Beaten Track …', *Geographical*, 76(7): 96.

Hall, L. (2005) 'The Long Way Home: A History of Motor Touring in Australia, 1925–2004'. Ph.D. thesis, University of Sydney.

Hall, L. (2006) 'The "Zest of Adventure" in Australian Overland Narratives, 1920–2000', *Journal of Tourism and Cultural Change*, 4(2): 85–95.

Ham, P. (2004) *Kokoda*. Sydney: HarperCollins.

Hampton, M.P. (2013) *Backpacker Tourism and Economic Development: Perspectives from the Less Developed World*. London and New York: Routledge.

Hatfield, W. (1936) *Australia through the Windscreen*. Sydney: Angus & Robertson.

Haynes, R. (1998) *Seeking the Centre: The Australian Desert in Literature, Art and Film*. Cambridge: Cambridge University Press.

Hirst, J.B. (1978) 'The Pioneer Legend', *Historical Studies*, 18(71): 316–337.

Hubbard, T. (2014) 'The Kangaroo Route: What Significant Fabric Survives?', *Historic Environment*, 26(1): 76–87.

Hyland, P. (2009) *Space Cake in Chicken Street: Travelling the Hippie Trail London-to- Kathmandu 1977*. Geelong: the author.

Jakle, J.A. and Sculle, K.A. (2008) *Motoring: The Highway Experience in America*. Athens: University of Georgia Press.

Jeans, P. (1998) *Long Road to London*. Perth: Rawlhouse Publishing.

Jensen, E. (2009) 'Kokoda Track Still a Testing Ground', *Sydney Morning Herald*, 17 October: 3.

Johnston, A. and Lawson, A. (2000) 'Settler Colonies'. In: H. Schwartz and S. Ray (eds) *A Companion to Postcolonial Studies*. Malden, MA: Blackwell Publishers, pp. 360–377.

Kerouac, J. (1972) [1957] *On the Road*. London: Penguin.

Kerr, R. (2012) 'On "the Road": A Cultural History: Imagining, Experiencing and Representing "the Road" in Australia 1890–2011'. Ph.D. thesis, University of Sydney.

Kerr, R. (2013) 'Through the Rear View Mirror: Landscapes, Legends and Literature on the Australian Road', *Studies in Travel Writing*, 17(2). Online. Available at: www.tandfonline.com/doi/abs/10.1080/13645145.2013.783960. Accessed 19 March 2018.

Kerr, R. (Forthcoming) *On the Road in Australia: A Cultural History*. Bristol: Channel View Publications.

Laing, A. (1975) *Eurasian Moke: The Story of a Journey*. Melbourne: Hill of Content.

Laing, J. and Frost, W. (2014) *Explorer Travellers and Adventure Tourism*. Bristol: Channel View Publications.

Lee, R. (2010) *Transport: An Australian History.* Sydney: UNSW Press.
Leed, E.J. (1991) *The Mind of the Traveller: From Gilgamesh to Global Tourism.* New York: Basic Books.
Limerick, P.N. (1994) 'The Adventures of the Frontier in the Twentieth Century'. In: J. R. Grossman (ed.) *The Frontier in American Culture: Essays by Richard White and Patricia Nelson Limerick.* Berkeley: University of California Press, pp. 67–95.
Lockwood, D. (1964) *Up the Track.* Adelaide: Rigby.
Macintyre, S. and Clark, A. (2004) *The History Wars.* Carlton: Melbourne University Press.
Mackay, D. (1976) 'Buddhism on a Queensland Mountain', *Australian Women's Weekly*, 18 August: 36–39.
Maclean, R. (2006) *Magic Bus: On the Hippie Trail from Istanbul to India.* London: Viking.
Novelli, M. (ed.) (2005) *Niche Tourism: Contemporary Issues, Trends and Cases.* Amsterdam and Sydney: Elsevier.
O'Brien, S. (producer) (2012) 'Asia Overland: Adventures on the Hippie Trail', *Hindsight*, ABC Radio National, 8 April. Online. Available at: www.abc.net.au/radionational/programs/hindsight/asia-overland/5155196. Accessed 1 September 2013.
Olding, R. (2012) 'Blood, Tears and Sweat', *Sydney Morning Herald*, 21 January: 10–11.
Parkinson, J. (2001) *Farewell Hippie Heaven: Rites of Way on the Overland Route.* Port Melbourne: Lothian Books.
Paterson, A.B. (1917) *Saltbush Bill, J.P. and Other Verses.* Sydney: Angus & Robertson.
Patton, P. (1986) *Open Road: A Celebration of the American Highway.* New York: Simon & Schuster.
Picard, D. and Robinson, M. (eds) (2012) *Emotion in Motion: Tourism, Affect and Transformation.* Farnham: Ashgate.
Primeau, R. (1996) *Romance of the Road: The Literature of the American Highway.* Bowling Green, OH: Bowling Green State University Popular Press.
Rees, C. and Rees, L. (1953) *Spinifex Walkabout: Hitch-hiking in Remote North Australia.* Sydney and London: Australasian Publishing Company in association with-George G. Harrap & Co.
Reynolds, H. (1998) *This Whispering in Our Hearts.* St Leonards: Allen & Unwin.
Robinson, M. (2012) 'The Emotional Tourist'. In D. Picard and M. Robinson (eds) *Emotion in Motion: Tourism, Affect and Transformation.* Farnham: Ashgate, pp. 21–48.
Scates, B. (2013) *Anzac Journeys: Returning to the Battlefields of World War II.* Cambridge: Cambridge University Press.
Shaffer, M.S. (2001) *See America First: Tourism and National Identity, 1880–1940.* Washington, DC: Smithsonian Institution Press.
Smith, L. (2012) 'A Pilgrimage of Masculinity: The Stockman's Hall of Fame and Outback Heritage Centre', *Australian Historical Studies*, 43(3): 472–482.
Sobocinska, A. (2006) 'The Ultimate Trip', *News National Library of Australia*, August: 8–11.
Tamworth Four Wheel Drive Club (1987) *Visitors' Log, Poepples Corner, Simpson Desert 1978–87.* MS 9123. Canberra: National Library of Australia.
Taylor, S., Varley, P. and Johnston, T. (eds) (2013) *Adventure Tourism: Meaning, Experience and Learning.* New York: Routledge.

Tomory, D. (1998) *A Season in Heaven: True Tales from the Road to Kathmandu*. Melbourne: Lonely Planet.
Tourism Australia (2017) 'The Great Australian Outback Cattle Drive, South Australia'. Online. Available at: www.australia.com/en–in/places/sa/outback–cattle–drive.html. Accessed 22 February 2017.
Veracini, L. (2010) *Settler Colonialism: A Theoretical Overview*. Basingstoke: Palgrave Macmillan.
Ward, R. (1977) [1958] *The Australian Legend*. Melbourne: Oxford University Press.
Ward, R. (1978) 'The Australian Legend Re-visited', *Historical Studies*, 18(71): 171–190.
Waterhouse, R. (2005) *The Vision Splendid: A Social and Cultural History of Rural Australia*. Fremantle: Curtin University Books.
Wheeler, T. (1973) *Across Asia on the Cheap: A Complete Guide to Making the Overland Trip*. Sydney: Lonely Planet.
Wheeler, T. and Wheeler, M. (2005) *Once while Travelling: The Lonely Planet Story*. Camberwell, Vic: Penguin.
White, R. (1992) *Inventing Australia*. Sydney: Allen & Unwin.
White, R. (1997) 'The Retreat from Adventure: Popular Travel Writing in the 1950s', *Australian Historical Studies*, 28(109): 90–105.
White, R. (2005) *On Holidays: A History of Getting away in Australia*. North Melbourne: Pluto Press Australia.
White, R. (2009) 'Australian Odysseys: Modern Myths of Travel'. In: A. Sarwal and R. Sarwal (eds) *Reading Down Under: Australian Literary Studies Reader*. New Delhi: SSS Publications, pp. 409–418.
Winter, J. (2006) *Remembering War: The Great War between Memory and History in the Twentieth Century*. New Haven, CT, and London: Yale University Press.
Woodburn, K. (1947) 'The Bitumen', *Walkabout*, March: 16–17.

Index

Note: Entries in **bold** refer to Figures.

For Product Safety Concerns and Information please contact our EU representative GPSR@taylorandfrancis.com
Taylor & Francis Verlag GmbH, Kaufingerstraße 24, 80331 München, Germany

www.ingramcontent.com/pod-product-compliance
Ingram Content Group UK Ltd.
Pitfield, Milton Keynes, MK11 3LW, UK
UKHW021303190926
475701UK00009B/625
9780367589189